NEXT-GENERATION DNA SEQUENCING INFORMATICS

OTHER RELATED TITLES FROM COLD SPRING HARBOR LABORATORY PRESS

A Short Guide to the Human Genome
Guide to the Human Genome
Molecular Cloning: A Laboratory Manual, Fourth Edition

HANDBOOKS

An A to Z of DNA Science: What Scientists Mean When They Talk about Genes and Genomes
At the Bench: A Laboratory Navigator, Updated Edition
At the Helm: Leading Your Laboratory, Second Edition
An Illustrated Chinese–English Guide for Biomedical Scientists
Binding and Kinetics for Molecular Biologists
Career Opportunities in Biotechnology and Drug Development
C. elegans *Atlas*
Experimental Design for Biologists
Fly Pushing: The Theory and Practice of Drosophila *Genetics*, Second Edition
Is It in Your Genes? The Influence of Genes on Common Disorders and Diseases That Affect You and Your Family
Lab Dynamics: Management and Leadership Skills for Scientists, Second Edition
Lab Math: A Handbook of Measurements, Calculations, and Other Quantitative Skills for Use at the Bench
Lab Ref, Volume 1: A Handbook of Recipes, Reagents, and Other Reference Tools for Use at the Bench
Lab Ref, Volume 2: A Handbook of Recipes, Reagents, and Other Reference Tools for Use at the Bench
Statistics at the Bench: A Step-by-Step Handbook for Biologists

NEXT-GENERATION DNA SEQUENCING INFORMATICS

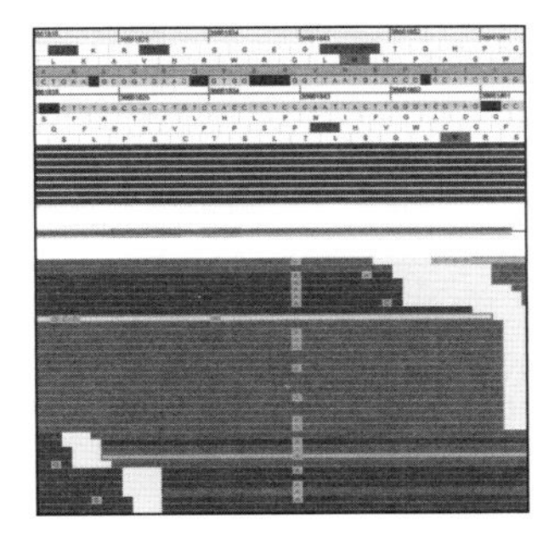

EDITED BY

STUART M. BROWN

COLD SPRING HARBOR LABORATORY PRESS
Cold Spring Harbor, New York • www.cshlpress.org

NEXT-GENERATION DNA SEQUENCING INFORMATICS

Printed in the United States of America

Publisher and Acquisition Editor	John Inglis
Director of Editorial Development	Jan Argentine
Project Manager	Inez Sialiano
Permissions Coordinator	Carol Brown
Production Editor	Kathleen Bubbeo
Production Manager	Denise Weiss
Sales Account Manager	Elizabeth Powers
Cover Designer	Michael Albano

Front cover artwork: A heterozygous single-nucleotide G>A variant is verified by visualization in Genome View (genomeview.org) of short reads from a next-generation sequencing machine aligned to the reference genome. Forward reads are shown in blue, reverse reads are shown in green, and sequence variants are highlighted in yellow. Other visible sequence variants are probable sequencing errors.

Library of Congress Cataloging-in-Publication Data

Next-generation DNA sequencing informatics / edited by Stuart M. Brown.

pages cm
Includes bibliographical references and index.
ISBN 978-1-936113-87-3 (hardcover : alk. paper)
1. Nucleotide sequence. 2. Bioinformatics. I. Brown, Stuart M., 1962-

QP625.N89N485 2013
572.8'633–dc23

2012034431

10 9 8 7 6 5 4 3 2

All World Wide Web addresses are accurate to the best of our knowledge at the time of printing.

For a complete catalog of all Cold Spring Harbor Laboratory Press publications, visit our website at www.cshlpress.org.

Contents

Preface

Next-generation DNA sequencing (NGS) technology has been a huge stimulus for new and exciting ways to create and test new hypotheses in biology as well as to revisit old ones but with a novel and vastly enhanced perspective. It would be no exaggeration to state that many of the current dynamic advances in biomedical basic and translational science are being driven by this technology.

NGS is enabled by sophisticated and novel bioinformatics tools specifically created or adapted to make NGS possible. Not only has new software been developed for a wide range of novel applications and types of data analysis, but new algorithms have also been developed for old problems, such as sequence alignment and de novo assembly, to cope with the huge volume of data generated on new sequencing machines.

The cycle of software development has accelerated as vendors upgrade their machines and different groups compete to publish new methods and to meet investigator demands. As a result of the frenetic pace of development, new software tools for NGS data analysis are often released with bare bones command line user interfaces and minimal documentation. Making things even more complicated, many different software packages exist for each of the major NGS applications with few benchmarking studies available to guide users in the choice of the best solutions. In short, there is an urgent need for a scientifically rigorous, cutting-edge, and practical treatise to guide researchers about all major aspects of informatics needed to successfully operate and fully take advantage of NGS.

The authors of the present work have been very lucky that their home institution, NYU Langone Medical Center, has invested early and heavily in building both assay and informatics capacity and manpower in NGS. Specifically, in 2008 NYU Langone Medical Center built its Genome Technology Center to provide research and translational scientists access to the latest DNA sequencing, expanding upon previous technologies such as microarrays and real-time polymerase chain reaction (qPCR). In parallel, the Informatics Center at NYU Langone Medical Center has developed the Sequencing Informatics Group to provide research design, upstream data processing, data management, and data analysis consulting for all users of the sequencers within NYU Langone Medical Center and beyond.

As our group has grown in experience, we have evaluated many different software packages and built best practice workflows for many different types of NGS projects, including de novo sequencing (and genome annotation), amplicon sequencing for rare variant detection and for metagenomics, ChIP-seq, RNA-seq, and detection of somatic variants in cancer (including single-base substitutions, insertions, deletions, and translocations).

In this book, building on our own extensive experience that spans collaborations on more than 30 National Institutes of Health–funded projects, and by critically evaluating and synthesizing the literature in the field, we provide an overview of many core types of NGS projects, a discussion of methods embodied in popular software, and detailed descriptions of our own best practice workflows (including several tutorials). We have included advice designed to be helpful to both bioinformaticians implementing their own data analysis methods and to laboratory and clinical investigators planning to use NGS methods to address their own research questions.

The future of NGS and all the related informatics innovations is as bright as it is exciting, and we are gratified to be able to contribute to the field's development with the present volume.

STUART M. BROWN

Acknowledgments

The authors would like to acknowledge the senior leadership of the NYU Langone Medical Center, Dean and CEO Dr. Robert Grossman and the entire executive and scientific leadership team, for creating an exceptionally enabling environment in which we were given the means and the encouragement to pursue our scientific investigations of NGS informatics.

We are also immensely grateful to all our colleagues at NYU Langone Medical Center who have entrusted us with the success of their NGS projects and allowed to us to create, test, and deploy a plethora of innovative informatics solutions across a diverse spectrum of basic science and translational investigations of the highest quality.

We are deeply indebted to John Inglis at Cold Spring Harbor Laboratory Press for seeing the value of this book. We are thankful for the great patience of the CSHLP staff in working with our erratic schedules and the superb quality they bring to every phase of production. We especially thank Inez Sialiano for her editorial guidance in all phases of the writing and Kathleen Bubbeo for insisting on the highest quality graphics and for rooting out errors throughout the text.

The NYU Sequencing Informatics Group:

Alexander Alekseyenko
Constantin Aliferis
Silvia Argimón (affiliate)
Stuart M. Brown
Efstratios Efstathiadis
Frank Hsu (affiliate)
Kranti Konganti
Eric R. Peskin
Christina Schweigert (affiliate)
Steven Shen
Phillip Ross Smith
Alexander Statnikov
Zuojian Tang
Jinhua Wang

About the Authors

Alexander Alekseyenko is Assistant Professor in the Department of Medicine and Associate Operations Director of the Bioinformatics consulting group for the Center for Health Informatics and Bioinformatics, NYU School of Medicine. Dr. Alekseyenko received his Ph.D. in Biomathematics from the University of California at Los Angeles. He conducted postdoctoral training first at the European Bioinformatics Institute, Cambridge, United Kingdom and then at Stanford University. Dr. Alekseyenko is the primary informatics faculty member at NYU working in the area of metagenomics, understanding the diversity of microorganisms present in the human body through next-generation sequencing and the application of evolutionary and ecological statistical models.

Silvia Argimón is Associate Research Scientist in the Cariology and Comprehensive Care Department at NYU College of Dentistry. Her research interests include oral bacteria diversity and virulence. She received her Ph.D. in molecular biology from University of Aberdeen, Scotland.

Stuart M. Brown is Associate Professor in the Cell Biology Department and a senior faculty member in the Center for Health Informatics and Bioinformatics at NYU School of Medicine, where he serves as Operations Director for the Bioinformatics consulting group and leader of the Sequence Informatics group. He has taught graduate courses in Bioinformatics at NYU for 12 years and he is the author of textbooks on bioinformatics and medical genomics. He received his Ph.D. in molecular biology from Cornell University.

Efstratios Efstathiadis is Assistant Professor and the Technical Director of the High Performance Computing Facility of the NYU Langone Medical Center. Previously he served as Technology Architect of the Center for Computational Science at Brookhaven National Laboratory. Dr. Efstathiadis obtained his Ph.D. in nuclear physics in 1996 at the City University of New York.

Jeremy Goecks is a Postdoctoral Fellow in the Departments of Biology and Math & Computer Science at Emory University. He is a core member of the team developing Galaxy, a popular Web-based platform for computational biomedical research. Dr. Goecks earned his Ph.D. in computer science from the Georgia Institute of Technology.

D. Frank Hsu is Clavius Distinguished Professor of Science and Professor of Computer and Information Science at Fordham University. He is former chair of the Fordham Computer Science Department. Dr. Hsu is the former Editor-in-Chief of the *Journal of Interconnection Networks*. He received his Ph.D. from the University of Michigan.

Kranti Konganti is a programmer/bioinformatician in the Center for Health Informatics and Bioinformatics at NYU School of Medicine. He received his M.S. in Bioinformatics from Northeastern University. He has primary responsibility at NYU for data analysis of sequencing performed on the Roche 454 machine as well as genome data visualization in the GBrowse system.

Eric R. Peskin is Associate Technical Director of the High-Performance Computing Facility at the Center for Health Informatics and Bioinformatics, NYU School of Medicine. Dr. Peskin earned his Ph.D. in computer science from the University of Utah. Previously, he served at Intel as a Senior Software Engineer in logic technology development and as Assistant Professor of Electrical Engineering at the Rochester Institute of Technology.

Christina Schweikert is Assistant Professor in the Computer and Information Science Department at Fordham University. Dr. Schweikert obtained her Ph.D. in computer science from the City University of New York.

Steven Shen is Associate Professor in the Department of Biochemistry and the Center for Health Informatics and Bioinformatics at NYU School of Medicine. The primary focus of his work is to develop next-generation sequencing–related technology and computational methods for probing the epigenetic alteration in the genomes of ant species. Before coming to NYU School of Medicine, Dr. Shen was Assistant Professor at Boston University School of Medicine and Research Scientist at Massachusetts Institute of Technology. He also worked at Helicos Biosciences developing single-molecule sequencing technology.

Phillip Ross Smith is Associate Professor in the Cell Biology Department and a senior faculty member in the Center for Health Informatics and Bioinformatics at NYU School of Medicine. He is a former CIO of NYU School of Medicine and a former editor of the *Journal of Structural Biology*. Dr. Smith obtained his Ph.D. in high energy physics from the University of Cambridge, United Kingdom and his M.D. from NYU School of Medicine.

Zuojian Tang is Associate Research Scientist at the Center for Health Informatics and Bioinformatics at NYU School of Medicine. She manages computing support for Illumina Next-Generation Sequencing. She received her M.S. in computer science and bioinformatics from McGill University.

James Taylor is Assistant Professor in the Departments of Biology and Mathematics & Computer Science at Emory University. He is one of the original developers of Galaxy, a popular Web-based platform for computational biomedical research. Dr. Taylor received his Ph.D. in computer science from Pennsylvania State University, where he was involved in several vertebrate genome projects and the ENCODE project.

Jinhua Wang is Assistant Professor at NYU School of Medicine and a member of the NYU Cancer Institute. Dr. Wang completed his Ph.D. training in computational biology and genomics at the Chinese Academy of Sciences. He also served as bioinformatics research manager for the Chinese National Human Genome Center. He conducted postdoctoral research at Cold Spring Harbor Laboratory, where he focused on developing mathematical and statistical methods to identify functional elements in eukaryotic genomes, especially on sequence elements that regulate gene transcription and pre-mRNA splicing. He also served as bioinformatics scientist at St. Jude Children's Research Hospital.

1

Introduction to DNA Sequencing

Stuart M. Brown

HISTORY OF DNA SEQUENCING

All of the DNA sequencing work for the **Human Genome Project** (1995–2003) was performed using modifications of the method invented by Frederick Sanger in 1975 (Sanger and Coulson 1975). Before Sanger's work, some nucleotide sequences were determined using ad hoc methods that involved RNA synthesis and enzymatic digestion. An interesting approximation of the Sanger method was published in 1971 by Ray Wu of Cornell University (Wu and Taylor 1971), where he was able to determine the 12-base single-stranded ends of bacteriophage λ DNA by the addition of complementary radiolabeled nucleotides to the single strand by DNA polymerase, followed by a complex scheme of nuclease digestion and chromatography. Walter Gilbert and Allan Maxam published a 24-bp sequence of the *lac* operator (a transcription repressor binding site) from the *Escherichia coli* genome in 1973 (Gilbert and Maxam 1973). Their method involved a complicated mixture of pyrimidine fingerprinting by partial nuclease digestion and chromatography as well as nuclease digestion of in vitro–transcribed RNA molecules. The entire sequence of the coat protein gene from bacteriophage MS2 was determined by Walter Fiers and coworkers at the University of Ghent, Belgium (Min Jou et al. 1972). This method relied on nuclease digestion of phage RNA, partial in vitro synthesis of RNA by RNA polymerase and incomplete nucleotide mixtures, and chemical characterization of the fragments. But Fiers also used information from the known protein sequence to limit the possible codons and to assemble overlapping fragments.

The sequencing method developed by Sanger in 1975 relies on the synthesis of new **DNA fragments** using DNA polymerase to extend a short synthetic oligonucleotide primer hybridized to a single-stranded DNA template. The first version of Sanger's sequencing method used a two-phase DNA synthesis reaction. In the first phase, the **sequencing primer** was partially extended using a mixture of all four deoxyribonucleotide triphosphates (dATP, dCTP, dGTP, and dTTP), generating a set of newly synthesized DNA fragments, all starting at the primer but extending for "random" lengths. In the second phase, the partially extended templates were split into four parallel DNA synthesis reactions, each one including only three of the four deoxyribonucleotide triphosphates. "Synthesis then proceeds as far as it can on each chain: thus, if dATP is the missing triphosphate, each chain will terminate at its 3′ end at a position before an A residue" (Sanger and Coulson 1975). The newly synthesized DNA fragments were then denatured from the template and separated by size by electrophoresis in adjacent lanes of an acrylamide gel. "Ideally, the sequence of the DNA is read off from the radioautograph" (Sanger and Coulson 1975) (see Fig. 1).

The **Sanger sequencing method** was revolutionary in several ways. Most importantly, it could be applied to any DNA molecule, and it could be used to determine long DNA sequences. However, this system, as first presented, had two critical limitations that prevented its immediate widespread adoption. First, the requirement for an oligonucleotide primer means that some DNA sequence must be known at a

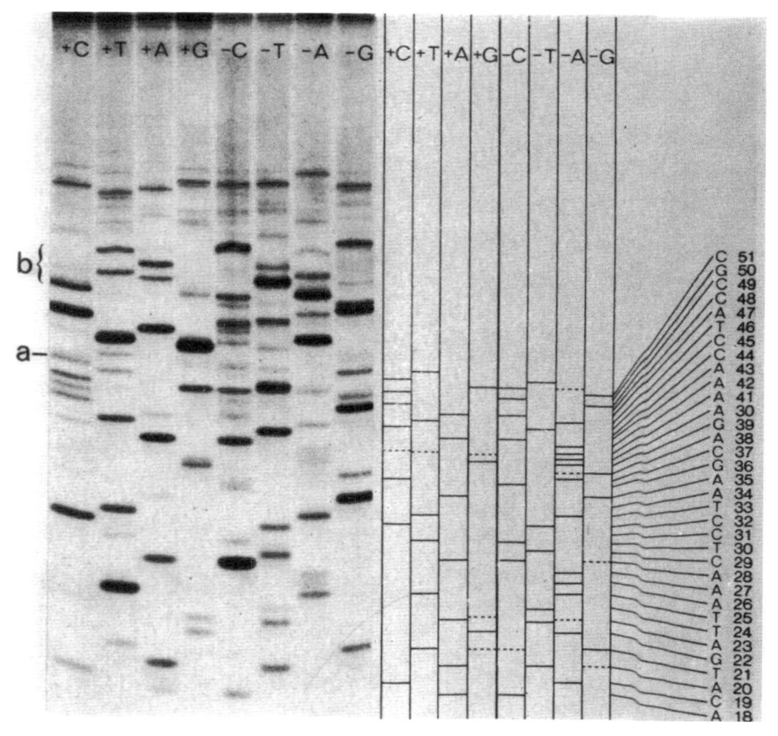

FIGURE 1. The autoradiograph produced by Sanger and Coulson in the 1975 *Journal of Molecular Biology* (**94:** 441–448) paper to document their method for DNA sequencing "by Primed Synthesis with DNA Polymerase."

location directly adjacent to the region of DNA where the sequence is to be determined. Second, the "random extension" of the primer does not necessarily generate an even distribution of fragments of all desired lengths.

Shortly after Sanger's "primer extension" sequencing method was published, Allan Maxam and Walter Gilbert invented a sequencing method based on chemical cleavage of DNA (Maxam and Gilbert 1977). Like the Sanger method, Maxam–Gilbert sequencing splits the DNA template into four reactions. In each reaction, the template is radioactively labeled at the 5′ end, then subjected to chemicals that specifically cleave DNA at one of the four bases. The reactions are conducted under conditions that produce, on average, just one cleavage per DNA molecule, at a random location. Then, like the Sanger method, the four reactions are loaded into adjacent lanes of an acrylamide gel, and fragments are separated by size by electrophoresis. The DNA sequence can then be read from an autoradiograph of the acrylamide gel. Maxam and Gilbert sequencing was initially more popular than Sanger sequencing because it can be conducted directly on purified DNA fragments, with no requirement for a single-stranded template and a complementary oligonucleotide primer.

Sanger rapidly improved his method by using dideoxynucleotides as "chain terminators" in the primer extension reaction in place of the clumsy two-phase procedure described in 1975 (Sanger et al. 1977). The improved method again starts with a single-stranded DNA template hybridized with a short complementary oligonucleotide primer. The primed template is split into four reaction mixtures, each containing DNA polymerase, the four normal deoxyribonucleotide triphosphates (with one radiolabeled nucleotide), and one dideoxynucleotide. As the polymerase extends the primer, whenever a dideoxynucleotide is incorporated, the reaction stops, producing a mixture of truncated fragments of varying lengths, all starting at the same primer and ending with the same base. Once again, the four reactions are loaded onto an acrylamide gel, the fragments are separated by electrophoresis, and the DNA sequence is read off the autoradiograph. Sanger reported reading sequences up to 300 bases long on a single gel.

Sanger sequencing and Maxam–Gilbert sequencing remained in competition for many years. The Sanger method became more popular, possibly because of the complexity of the steps and the toxicity of the reagents used in the Maxam–Gilbert method. Many refinements have been developed to improve the Sanger method, including a variety of cloning methods for the preparation of single-stranded templates that span a gene (or an entire genome) of interest as well as commercial kits to streamline the preparation of reagents. One very significant improvement in the Sanger technology was the development of fluorescent dyes to replace radioactive labels on the newly synthesized DNA fragments (Smith et al. 1986). This led to the development of semiautomated DNA sequencers by Leroy Hood, Michael Hunkapiller, and others that were commercialized by Applied Biosystems

Inc. (ABI). Crucial innovations in the ABI sequencers included attaching the four different colors of dye labels to the four dideoxynucleotide chain terminators, so that fragments terminated at all four bases could be generated in a single reaction tube and assayed on a single lane of an acrylamide gel, and using a computer to monitor a real-time fluorescent detector, so that the sequence data could be collected automatically as the gel electrophoresis was run. These ABI sequencers provided nearly all of the data for the Human Genome Project. Another incremental improvement in the ABI automated fluorescent sequencers was the use of acrylamide gel in capillary tubes rather than in a large thin slab between two glass plates. This saved setup work for technicians in the sequencing laboratory, allowed for more consistent results from electrophoresis, allowed for increased speed of electrophoresis, and allowed the machines to be scaled up to run more samples simultaneously (see Fig. 2).

Cloning for Sequencing

The Sanger sequencing reaction uses a single-stranded DNA template, a short single-stranded oligonucleotide primer that is complementary to the template, DNA polymerase enzyme, and a mixture of chain-extension and chain-terminating nucleotides. The usual strategy to prepare DNA for sequencing involves **cloning** a target fragment of DNA into a plasmid vector that provides a cloning site between binding sites for standard sequencing primers that can be used by single-strand DNA polymerase II. This allows any DNA target to be sequenced in both directions using standard oligonucleotide primers, so that the sequence of the target molecule does not need to be known in advance (see Fig. 3).

DNA sequencing using the Sanger method is capable of reading ~500–800 bases in a single read. This is a limitation of both the Sanger primer extension/chain-terminator chemistry and the ability to separate DNA fragments by electrophoresis with accurate single-base resolution. Because most interesting biological nucleic acid molecules (such as genes, mRNA transcripts, plasmids, and genomes) are much longer than 800 bp, DNA sequencing projects usually involve some strategy of breaking DNA molecules into shorter pieces, sequencing them, and then using bioinformatics tools to assemble the data into complete sequences of the target molecules.

For small sequencing targets, a strategy based on restriction digest fragments may be effective, although it may be difficult to keep track of the size and orientation of all of the fragments for assembly of the sequences. Another strategy developed by Henikoff (1984) involves the generation of progressively smaller fragments of DNA by directional digestion with exonuclease III. The nested sequences are then assembled by overlapping the reads to build a contiguous sequence (**contig**). Because all DNA sequencing methods generate some errors, it became standard procedure to combine several overlapping **sequence reads** over the entire extent of the target

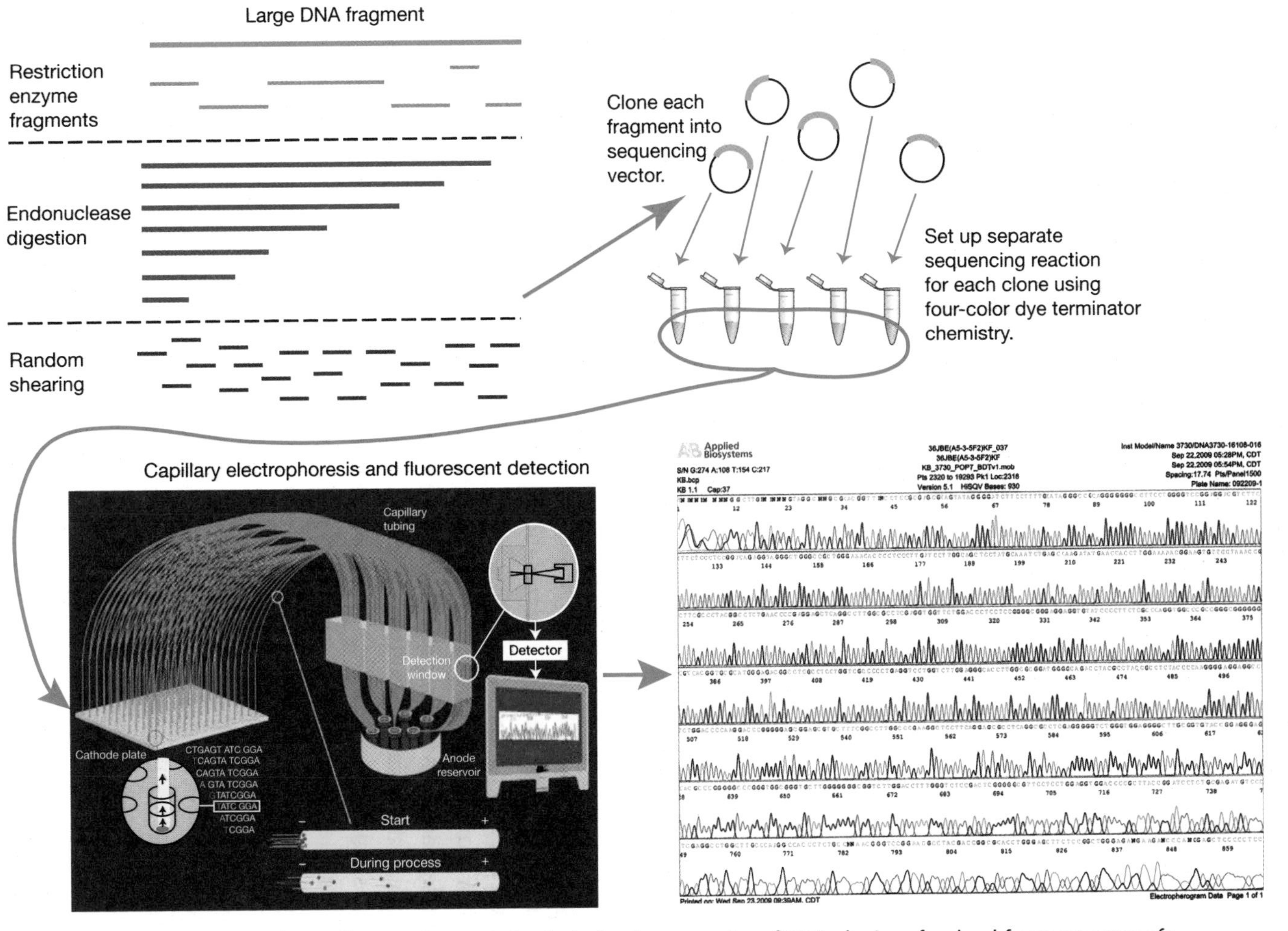

FIGURE 2. Workflow for **capillary DNA sequencing** including fragmentation of DNA, cloning of each subfragment, setup of a single sequencing reaction for each fragment, and loading of each reaction into a single capillary fiber for electrophoresis and fluorescent detection of individual dye-labeled terminator bases.

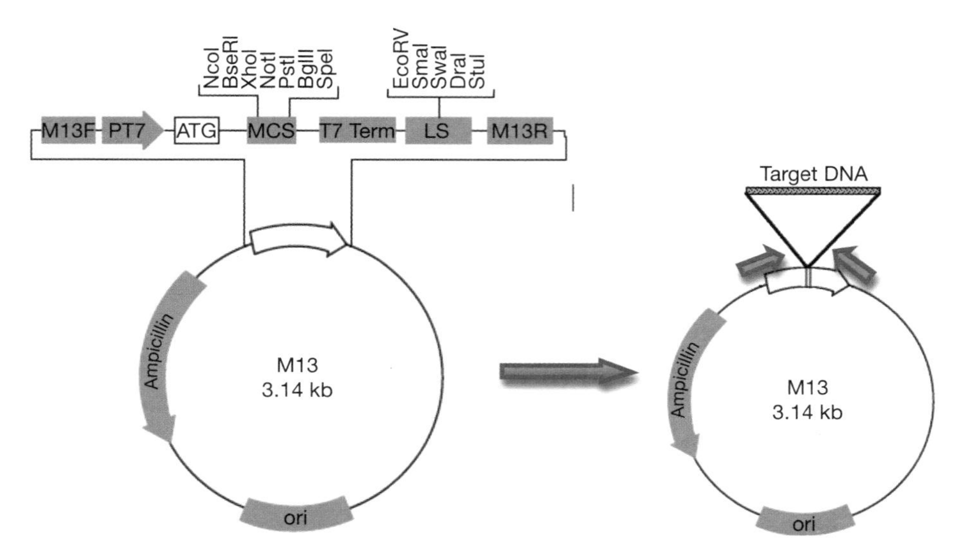

FIGURE 3. Cloning of a DNA fragment into the multiple cloning site of the M13 sequencing vector.

region, ideally reads from both directions on the DNA molecule. The process of assembling multiple overlapping reads in both directions into a **consensus sequence** became a focus for software development in the mid 1980s and 1990s. A review paper written in 1994 (Miller and Powell 1994) compared the performance of 11 different DNA **sequence assembly** programs.

Fragment assembly for sequencing projects is algorithmically similar to **sequence alignment**, but it has some unique aspects. First, because each sequence read is produced from a fragment cloned into a plasmid (or virus) vector, the first few bases of data often contain portions of the sequencing vector. In cases in which the length of the cloned fragment is smaller than the length of the sequence read, the end of the sequence read may also contain vector. It is also possible to create cloning artifacts where the entire sequenced region consists of only vector DNA. Therefore, it is necessary to identify and remove all vector sequences from raw sequence data before assembly of reads into contigs.

Second, the quality of DNA sequences obtained by the Sanger method is not consistent. The first ~50 bases are low quality because of uneven separation by electrophoresis and noise created by unincorporated primers and primer-dimer artifacts. The ends of sequence reads (beyond 500 bases) are low quality because of reduced signal (lower amounts of fragments of these sizes), diffusion, and poor separation of fragments due to smaller relative differences in electrophoretic mobility. Ideally, sequence assembly software should identify regions of low quality and provide tools to trim them away from good-quality sequence in the center of the read.

As sequence assembly software improved, new strategies were developed for ambitious sequencing projects. Instead of carefully cloning restriction fragments or nuclease-digested nested deletion fragments, investigators realized that DNA could be randomly sheared into a collection of unordered fragments. These fragments could then be cloned and sequenced and assembled by using software to find overlaps (see Chapter 5). The entire process became known as **shotgun sequencing** (Anderson et al. 1982). The shotgun **DNA fragments** form a **Poisson distribution** across the target DNA molecule. Therefore a large number of DNA fragments must be sequenced in order to build a complete contig with adequate **coverage** of every base in the original molecule. For a 10,000-base DNA target (10 kb), fragments totaling the equivalent of eight to 10 times the total length must be sequenced. Shotgun strategies became more attractive as the cost of sequencing each lane was reduced compared with the cost of laboratory work to carefully generate restriction fragments and nested deletion clones.

For very large sequencing projects (entire genomes), a divide-and-conquer strategy was often used. Large fragments of DNA, from 100,000 to 1 million bases, can be cloned into vectors known as bacterial artificial chromosomes (BACs). These fragments can then each be sequenced by the shotgun strategy.

Even with moderately high coverage (8× – 10×), shotgun assembly of overlapping **sequence fragments** usually leaves some gaps in the sequence of a targeted gene region. These gaps may be caused by random low-coverage areas resulting from the Poisson distribution of fragments, or they may be caused by sequence-specific effects that interfere with the cloning or sequencing process for some portion of the target region. Gaps may be filled by a "primer walking strategy," which relies on designing specific sequencing primers that can be used to start sequencing reactions at the ends of contigs and extend the sequence to cover the gap region. As each sequence read is added to the contig, new primers can be designed until overlap is achieved with another contig. A primer walk in the opposite direction provides double-stranded coverage. This strategy is very efficient in terms of the number of sequencing reactions required to cover a section of DNA, but is very time-consuming because each primer can only be designed and synthesized once data are known from the sequencing reaction using the previous primer (see Fig. 4).

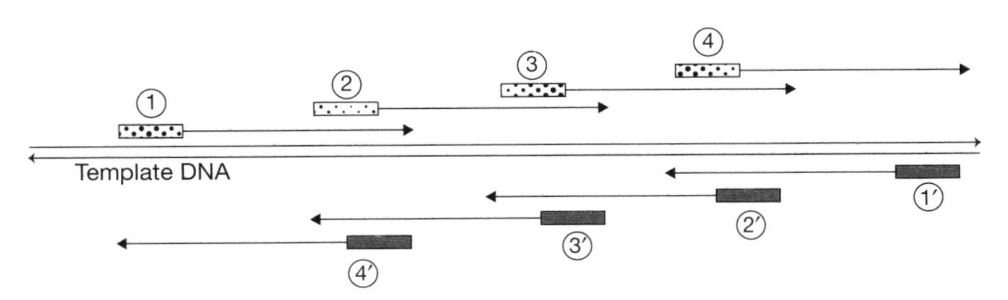

FIGURE 4. Primer walking strategy for sequencing.

NEXT-GENERATION SEQUENCING

It has been observed by several historians (Goldstein 1978; Kuhn 1996; Gladwell 2008) that when accumulated knowledge or new technologies become sufficient, a scientific problem may be simultaneously addressed by many different researchers, often producing simultaneous discoveries. The more successful strategies or theories then compete until one emerges as the most successful, which then becomes the standard method or the dominant paradigm. Clearly the 1970s was such a revolutionary period for DNA sequencing. Another such revolution in **next-generation DNA sequencing** (NGS) technology is currently underway. The stabilization of a new paradigm for DNA sequencing has clearly not yet occurred, because the output of sequencing machines has doubled while the cost per base has dropped by half in every year from 2004 to 2012. NGS technologies are generally characterized by very high data throughput, shorter sequence read length, and less accuracy compared with Sanger sequencing. Key bioinformatics challenges created by NGS data include either aligning (mapping) large numbers of reads to a **reference genome** or de novo assembly of novel genomes, **multiple alignment** of huge numbers of reads and rare **variant detection** for **amplicon** sequencing projects, and file formats and computational tools for efficient storage and manipulation of multigigabyte sequence data files.

Another interesting aspect of NGS is its effect on the location of sequencing technology within the scientific community. In the 1980s, most DNA sequencing was accomplished in small laboratories by hand pouring of polyacrylamide gels on large glass plates and painstaking reading of single bases from large X-ray films. As expensive automated high-throughput DNA sequencing machines became available, larger sequencing projects shifted to large specialized sequencing laboratories, core facilities, and dedicated DNA sequencing contractors. NGS may move DNA sequencing back into small laboratories if the NGS machines become cheaper, or expensive machines may create a permanent class of sequencing outsource vendors, which would function much like medical diagnostic laboratories. Even if NGS machines become very cheap or contract sequencing companies make NGS easily available to small laboratories, analysis of these large and complex data sets requires computational infrastructure and bioinformatics skills that are not found in most small clinical or molecular biology research laboratories.

Because many different vendors are producing high-throughput DNA sequencing machines that operate with significant differences in their underlying technology, as well as the types and quantities of sequence information produced, it is useful to define the term "Next-Generation DNA Sequencing" (NGS). In this book, *NGS* refers to DNA sequencing technologies that simultaneously determine the sequence of DNA bases from many thousands (or millions) of DNA templates in a single biochemical reaction volume. Each template molecule is affixed to a solid surface in a spatially separate location, and then clonally amplified to increase signal strength. The use of clonally amplified single-molecule templates allows for the detection and quantitation of rare variant sequences in a mixed population. These systems avoid the use of bacterial cloning steps, which can create sequencing bias or completely omit some regions of DNA. Other technologies are under development to sequence single long DNA templates very rapidly ("single molecule sequencing"), but this approach supports experiments with very different experimental designs and informatics implications, which are outside the scope of this book.

The internal workings of DNA sequencing machines are essentially engineering matters and involve proprietary technologies in areas such as fluidics, nanomaterials, chemistry, optics, and image processing. The informatics methods used to generate sequence data for DNA sequencing machines are proprietary to their manufacturers and under very rapid cycles of development and change (with no outside scientific review). Bioinformatics is generally defined as an open scientific discipline with peer-reviewed methods that analyzes molecular biology data such as DNA (and protein) sequences, gene expression values, genotypes, and the like, which can be produced by any type of assay technology. It might be assumed that "a sequence is a sequence" and the bioinformatician can proceed to analyze NGS data by de novo assembly, genome alignment, mutation detection, gene expression measurement, detection of protein binding, and other applications without regard to the equipment used to collect the data. However, the data produced by different types of DNA sequencing technologies are marked by technology-specific error patterns that can significantly influence downstream analysis. Each technology has a bias toward particular types of errors. For example, many of the NGS technologies are subject to various forms of read duplication due to PCR artifacts or optical artifacts, **pyrosequencing** tends to produce insertion and deletion errors in homopolymer sequences (runs of a single type of base), and Sanger sequencing is inaccurate in regions with high G + C content. Therefore, it is necessary to describe each of the currently common NGS technologies in some detail in order to highlight important informatics issues in the data that they produce.

454

The start of the NGS revolution is clearly marked by the publication in 2005 by 454 Life Sciences Corporation in *Nature* magazine of the complete sequencing of the *Mycoplasma genitalium* and *Streptococcus pneumoniae* genomes with 96% coverage at 99.96 % accuracy in one run of their novel parallel template pyrosequencing machine (Margulies et al. 2005). This was actually a follow-up to the public announcement (by press release) in 2003 that 454 had developed a novel "massively parallel" DNA sequencing technology with which they had sequenced the entire genome of adenovirus (30,000 bases) in a single day, including sample preparation (Pollack 2003). The initial Genome Sequencer 20 (GS20) product, offered commercially in 2005 by 454, produced ~25 million bases of high-quality DNA sequence per run, with reads 80–120 bases long. This represented a 10-fold improvement over the best Sanger-based system from Applied Biosystems, the 3730xl 96-Capillary Sequencer, which provides a maximum yield of 2.8 million bases per day (72 runs × 96 samples × 400 bp reads). The 454 method also eliminated the need to clone individual DNA fragments. In 2007, an improved version of the commercial 454 machine (GS FLX) was able to produce 100 million bases per run with reads 250 bases long. Using this improved system, Rothberg and colleagues published the complete genome sequence of James Watson, sequenced entirely on 454 machines (Wheeler 2008). In 2010, 454 released an upgrade to the GS FLX system known as Titanium, which increased the average read length to 400 bases for >1 million reads, for a total sequence yield of 400–600 million high-quality bases per run.

Sample preparation for the 454 system follows the shotgun strategy: random shearing of the genomic DNA, adding adapter sequences to the ends, then combining the DNA fragments with Sepharose beads (diameter ~28 μm) previously coated with oligonucleotides complementary to the adapters. The DNA is mixed with an excess of beads so that most beads bind only a single template molecule. Then the beads with bound DNA are subjected to emulsion PCR, which amplifies the DNA templates from a single copy to approximately 10 million copies on each bead. The enriched, template-carrying beads are deposited into open wells arranged along one face of a 60 × 60 mm^2 fiberoptic slide (picotiter plate). The wells are sized to fit only a single bead, and the plate contains approximately 2 million wells. Reagents are supplied to the picotiter plate for sequential rounds of **sequencing by synthesis** using a modification of the pyrosequencing method (Ronaghi et al. 1996). The location of each template molecule in its unique well of the 454 picotiter plate allows for the base to be recorded and allows computational assembly of the sequences of all templates to progress simultaneously. Although image processing methods are outside the scope of this discussion of sequencing informatics, it is noteworthy that the initial commercial release of the 454 GS20 sequencer was equipped with an integrated computer that contained a 6 million–gate FPGA coprocessor to allow for signal processing in real time.

Pyrosequencing was initially developed in 1996 by Nyrén and colleagues at the Royal Institute of Technology, Stockholm, Sweden (Ronaghi et al. 1996; Nyrén 2007), commercialized by Pyrosequencing AB, and licensed by 454 from QIAGEN. Like Sanger sequencing, pyrosequencing uses DNA polymerase to synthesize complementary strands to a single-stranded template, but it provides only one type of deoxynucleotide triphosphate base in a single cycle of the reaction. Each addition of a new nucleotide to a growing copy strand is accompanied by the release of pyrophosphate, which is converted to the emission of light by a pathway including ATP sylfurylase, luciferase, and luciferin. The chemiluminescent event is detected by a camera. Because only one type of nucleotide base is added during a cycle of DNA synthesis, the pyrosequencing chemistry has a very low rate of base call errors. However, the use of pyrosequencing chemistry creates one of the key drawbacks of the **454 sequencing** method. When a template molecule contains multiple bases of the same type, such as a run of AAAAs (a homopolymer), then multiple bases are synthesized onto the copy strand all at once, creating a larger emission of light. It is difficult for the system to count the number of bases accurately in homopolymers longer than 8 or 9 bases. Because many different template molecules are sequenced simultaneously in different wells of the picotiter plate and homopolymers of various length occur randomly, the length of the newly synthesized DNA copy strands will differ. DNA templates with many homopolymers will produce longer copy strands than those with sequences that contain only single bases. As a result, the 454 sequencing process produces a set of sequence reads with a distribution of different sizes (see Fig. 5).

454 Life Sciences claimed an early lead in the NGS field with its complete sequencing of the *M. genitalium* and *S. pneumoniae* genomes (Margulies et al. 2005). The 454 team followed up with several high-profile publications in collaboration with leading genomics scientists. In 2006, 454 collaborated with Svante Pääbo to produce 1 million bases of genomic DNA sequence from fossil Neanderthals (Green et al. 2006). The 454 system is particularly well suited to the study of ancient DNA because it does not require cloning; templates molecules are amplified and sequenced individually so that competition is minimized; and the production of a large number of **short reads** (100–200 nucleotides) matches the length of degraded DNA fragments that are recovered from fossils. 454 sequencing is also well suited to the "deep sequencing" of a specific target gene to find rare variants, such as somatic mutations within heterogeneous tumor samples (Thomas et al. 2005) or a low proportion of drug-resistant HIV **sequence variants** within the blood of a patient (Wang et al. 2007). 454 technology was used in more than 100 peer-reviewed publications by the end of 2007, including identification of a virus associated with honeybee colony collapse disorder (Cox-Foster et al. 2007), sequence of the mitochondrial DNA from wooly mammoth (Gilbert et al. 2007), and whole-genome sequencing of the Pinot Noir grape. The 454 system has also become popular for many **metagenomic** studies associated with the **Human Microbiome Project** (Nossa et al. 2010) (see Chapter 11).

454 machines produce data in a custom output file format known as .SFF (Standard Flowgram Format). This is a binary format, which contains base calls, **Phred** style quality scores for each base (see Chapter 2), and fluorescent intensity information. **SFF files** can contain sequences from a single sample, multiplex (barcoded) samples from a single region on the 454 plate, or data from multiple plate regions or multiple runs of the machine. SFF files are binary with nearly optimal data storage properties, thus file storage space is not saved by data compression software. The SFF

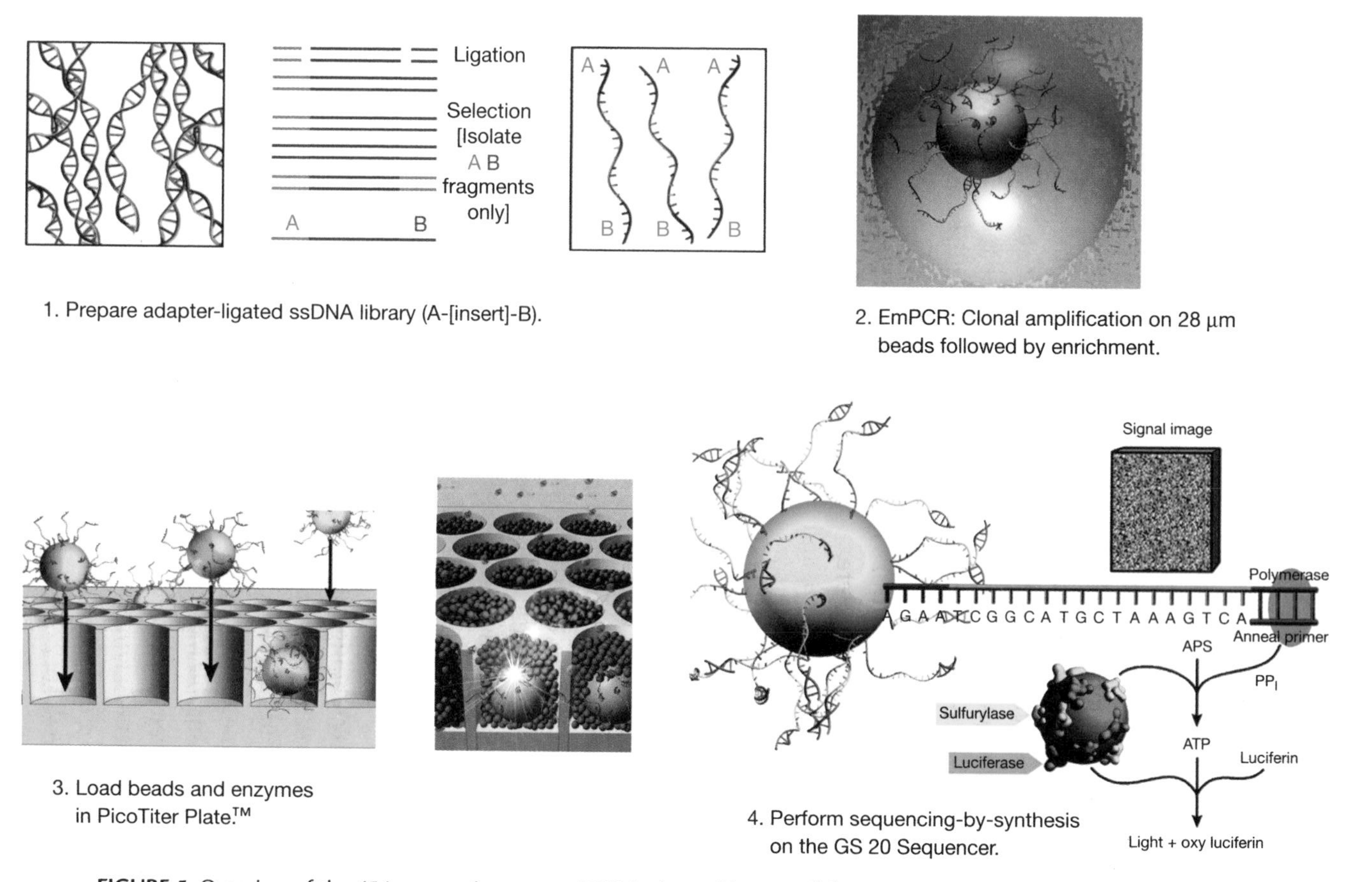

FIGURE 5. Overview of the 454 sequencing system. DNA is sheared into small fragments to which adapters are ligated, and fragments are attached to beads, mixed into an emulsion, amplified by emulsion PCR, and deposited in wells of a picotiter plate, then sequences are determined by pyrosequencing.

format was initially developed by 454, but a collaboration with the NCBI Trace Archive, Whitehead Institute for Biomedical Research, and the Sanger Institute has produced an open standard for this file type. See http://www.ncbi.nlm.nih.gov/Traces/trace.cgi?cmd=show&f=formats&m=doc&s=format#sff.

454 provides free software tools to read SFF files to its customers, but these tools are *not* available on the Internet. Flower is a free tool in the Haskell language (http://hackage.haskell.org/package/flower) to read SFF files and extract sequence data in **FASTA**, or sequence + quality in **FASTQ** format. Flowgram information can also be extracted as flow intensities (Malde 2011). SFF Workbench (http://www.dnabaser.com/download/SFF%20tools/index.html) is a free graphical tool to read SFF files developed by Heracle BioSoft S.R.L. as a companion to their DNA BASER software package. Both tools allow for the extraction of sequence reads (in FASTA format) and quality scores in a similar .qual format. The 454 machine does produce raw image data for each nucleotide flow in the sequencing process, but the (very large) image files are not generally stored for reanalysis after the initial QC process for each run of the machine is complete.

Illumina Genome Analyzer

The sequencing by synthesis technology used by the Illumina Genome Analyzer was originally developed by Shankar Balasubramanian and David Klenerman at the University of Cambridge (Furey et al. 1998). They founded the company Solexa in 1998 to commercialize their sequencing method. Solexa released its first commercial Genome Analyzer in 2006. The machine produced 1 billion bases of DNA sequence in a single run (~4 days). Illumina purchased Solexa in 2007. Illumina has produced very rapid upgrades in the Genome Analyzer technology. In 2011, the HiSeq 2000 produced 200 GB of data from a set of approximately 1 billion template molecules as 2 × 100-bp **paired-end reads**, at a rate of 25 GB per day. The HiSeq can also produce single end reads of 50 or 100 bp. In 2012, Illumina developed the MiSeq, a smaller single-lane sequencing machine. The MiSeq has a lower error rate, allowing for longer sequence reads. Illumina supported protocols in 2012 that provide up to 150-bp paired-end reads, but much longer reads have been reported by laboratories that have tested alternate protocols.

Conceptually, the Solexa/**Illumina sequencing** process is quite straightforward (see Fig. 6). The Sanger sequencing process copies a single-stranded DNA template with a DNA polymerase enzyme by adding a mixture of deoxyribonucleotide triphosphates (dNTPs) with a low concentration of chain-terminating dideoxynucleotides. The innovation of the Solexa method is to use modified dNTPs containing a terminator that blocks further polymerization, so that only a single base can be added to a growing DNA copy strand. In addition, the Solexa reaction is conducted on a very large number of different template molecules spread out on a solid surface.

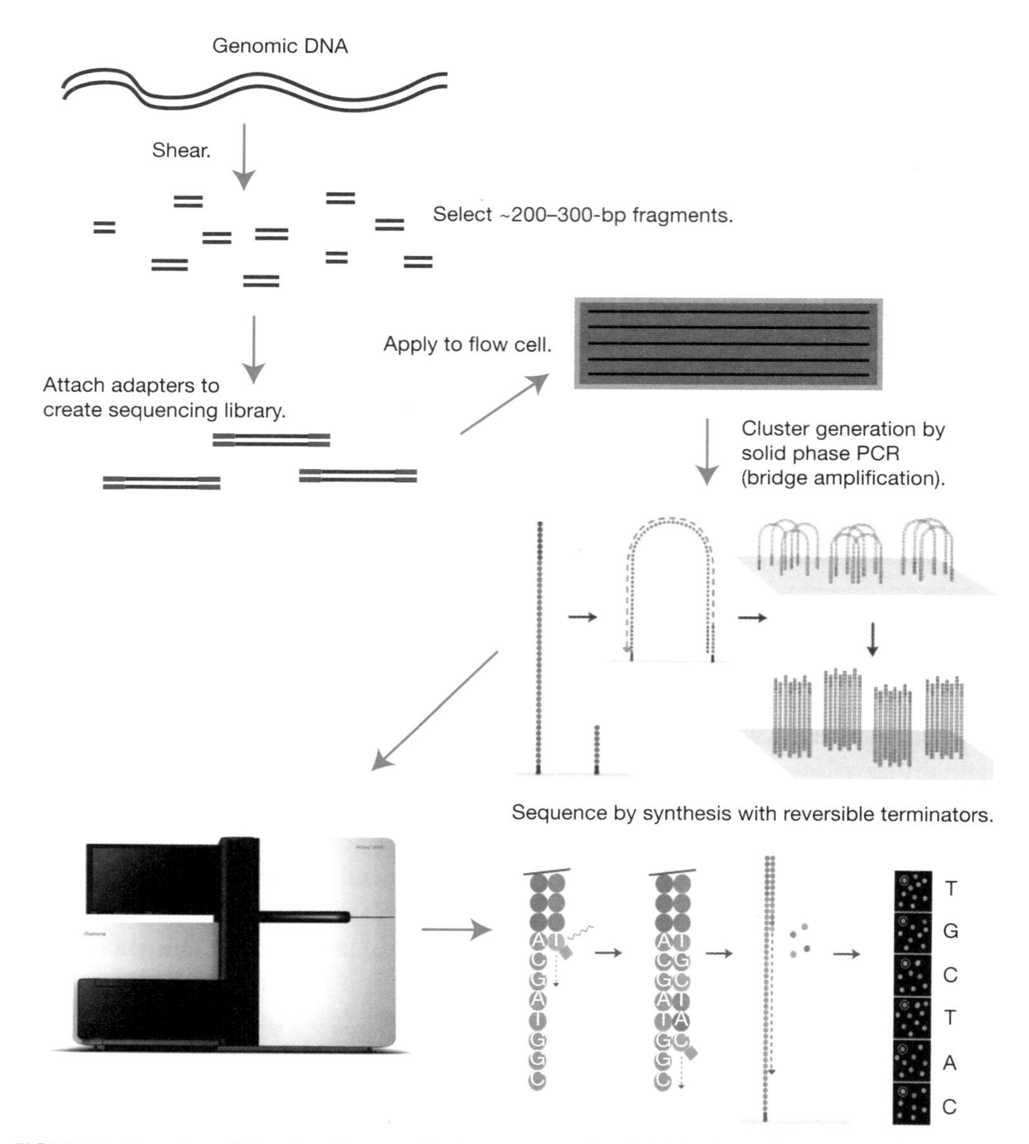

FIGURE 6. Overview of Illumina Genome Analyzer sequencing. DNA is sheared into small fragments to which adapters are ligated, fragments are attached to the surface of a flow cell, and in situ PCR "bridge" amplification creates clusters that are sequenced by reversible terminator chemistry.

The terminator also contains a fluorescent label that can be detected by a camera. The Solexa system uses only a single fluorescent color, thus each of the four dNTPs must be added in a separate cycle of DNA synthesis and imaging. After all four dNTPs have been added and the images recorded, the terminators are removed by an enzyme; thus the chemistry is called "reversible terminators." Then another four cycles of dNTP addition are initiated. Because single bases are added to all templates in a uniform fashion, the Solexa sequencing process produces a set of DNA sequence reads of uniform length, and the system is able to sequence homopolymer regions with no loss of accuracy. Illumina sequences typically have average error rates <1%.

The fluorescent imaging system used in the Genome Analyzer is not sensitive enough to detect the signal from a single template molecule. Another major innovation of the Solexa/Illumina method is the amplification of template molecules on a solid surface. The DNA sample is prepared into a "sequencing library" by fragmentation into pieces ~200 bases long, then custom adapters are added to each end. The library is then flowed across a solid surface (the "flow cell") that is coated with an oligonucleotide complementary to the adapter, so that the template fragments bind to the surface. Then a solid-phase "bridge amplification" process (cluster generation) creates approximately 1 million copies of each template in tight physical clusters on the flow-cell surface. One critical aspect of the clonal amplification on the flow-cell surface is that the locations of the clusters are randomly distributed, thus some clusters that derive from different single-molecule templates (with different sequences) will be located very close to each other, or even overlapping (see Fig. 7). The Illumina GA system relies on image analysis technology to resolve adjacent clusters. If adjacent clusters cannot be resolved, then data from both clusters will be discarded. Illumina has improved its image analysis technology dramatically since the first commercial release of the GA, allowing for much higher cluster density on the surface of the flow cell.

Solexa developed its sequencing technology without advertising its progress in scientific publications. However, soon after Illumina purchased the company, a series of papers appeared in high-impact journals highlighting the capabilities of the Illumina Genome Analyzer to produce unprecedented quantities of high-quality genomic sequence data. Three papers in the November 6, 2008 edition of *Nature* documented Illumina sequencing of entire human genomes, including one male

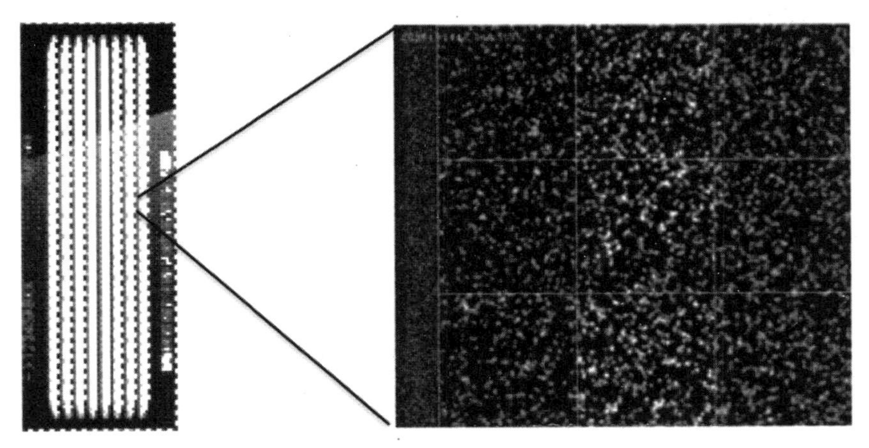

FIGURE 7. Image of an Illumina GA flow cell and a magnified view of a few grid squares on the cell showing fluorescent signal from individual clusters.

from Africa (Bentley et al. 2008), one male from China (Wang et al. 2008), and the genome of an acute myeloid leukemia (AML) tumor, as well as a matched normal skin sample from the same patient (Ley et al. 2008). For the African project, 4 billion paired-end 35-base reads were collected for a total of 135 GB of data (30× coverage of the human genome). These data were generated over a period of 8 weeks on six GA1 instruments, averaging 3.3 Gb per production run, with an estimated total cost of $250,000 for reagents. The Chinese individual was sequenced at the Beijing Genome Institute (using five Illumina GA sequencers) with 36× coverage of the human reference genome. For the AML project, the tumor genome was sequenced at 32× coverage and the matched normal genome at 14×. RNA sequencing on the Illumina GA was also introduced in a 2008 paper in *Science* magazine from the laboratories of Mark Gerstein and Michael Snyder at Yale University (Nagalakshmi et al. 2008).

All of these publications emphasized the ability of Illumina GA technology to produce significantly more genomic sequence data at lower cost than Applied Biosystems capillary Sanger-based machines or the 454 GS machine. It was also shown that for projects that involve resequencing of known genomes, short reads of 36 bases with high coverage could be used effectively to discover sequence polymorphisms and structural variants.

The data output of Illumina sequencers has gone through several very rapid changes, driven primarily by the huge increases in data. The first publicly released Genome Analyzer produced raw image data files as output, which required analysis by an external UNIX computing system. This created data transfer bottlenecks for the internal networks of many research institutions and created a substantial burden of IT support for each machine. The GA II generation of machines was able to process images during each cycle of sequencing on an integrated workstation, producing files of bases and quality scores in FASTQ format and making the storage of images optional. The 2011 HiSeq machine deletes images by default after image processing; however, external data processing is required to convert large numbers of files containing basecall and quality information for each run into standard formats such as FASTQ or BAM.

ABI SOLiD

Applied Biosystems (ABI, now a subdivision of Life Technologies Inc.), was the dominant supplier of DNA sequencing machines used in the Human Genome Project, and generally for all DNA sequencing work from the mid 1980s until approximately 2006. The early ABI sequencing machines used the Sanger sequencing chemistry and large polyacrylamide slab gels poured by hand between large sheets of glass. The key advantage of these machines over a completely homemade sequencing system was the four-color dye-terminator chemistry that allowed all four bases to be assayed simultaneously in a single reaction tube and a single gel lane. The system used fluorescent labels on the bases, which were detected by an automated data collection system (Smith et al. 1986), freeing technicians and graduate students from the task of manually reading bands on gels and manually typing letters representing the bases into text files. In 1998, ABI developed a new machine (the ABI 3700) with 96 capillary electrophoresis lanes, a significant improvement over the slab gels in throughput, automation, and accuracy. The ABI 3700 was used to produce most of the data for the Human Genome Project and all of the data produced by Celera Genomics Inc.

The commercial success of NGS machines such as the 454 Genome Sequencer and the Illumina Genome Analyzer challenged the dominance of ABI in DNA sequencing technology. In 2006, ABI acquired a novel sequencing technology called Supported Oligo Ligation Detection (SOLiD) by purchasing the biotech company Agencourt Personal Genomics. This **SOLiD sequencing** methodology is based on sequential ligation of dye-labeled oligonucleotide probes, whereby each probe assays two base positions at a time. ABI released SOLiD as a commercial DNA sequencing product in 2007, followed by rapid upgrades to SOLiD 2 in May 2008 and SOLiD 3 in October 2008. The overall sequencing output of the SOLiD is similar to the Illumina Genome Analyzer in terms of throughput and cost, but the chemistry of the sequencing technology is very different, which has a significant impact on bioinformatics methods used for data analysis.

Sample preparation for the SOLiD system is quite similar to that for 454. Genomic DNA is sheared into small fragments (~200 bp) and oligonucleotide adapters are added to both ends. Single-molecule templates are attached to magnetic beads coated with complementary oligonucleotides. Template molecules are amplified on the bead surface using emulsion PCR. The template-coated beads are then deposited on the surface of a glass slide, which is loaded into a flow cell. The sequencing template is made single stranded, and a sequencing oligonucleotide primer, which matches the adapter, is added. At this point SOLiD uses a unique DNA sequencing chemistry based on ligation rather than the DNA polymerase-based chemistry used in Sanger, 454, and Illumina systems.

The SOLiD ligation system uses a mixture of 16 fluorescently labeled oligonucleotides, each of which contain two specific bases at the 3′ end followed by three nonspecific bases (degenerate mixtures of the four bases). One of these oligos will hybridize as a complementary match to every possible 2-base template sequence following the sequencing primer. The oligo is then ligated to the primer. The ligation step is essential for specific detection of bases immediately 3′ of the primer, because oligos can hybridize to complementary sequences anywhere on the template. Unligated oligos are washed away. Then the oligo at each bead on the flow cell is identified by its fluorescent label. Note that because there are only four fluorescent labels, the detection identifies a set of four 2-base oligos, any one of which may have hybridized on a specific template/bead. The fluorescent label is then removed along with a terminator sequence (shown

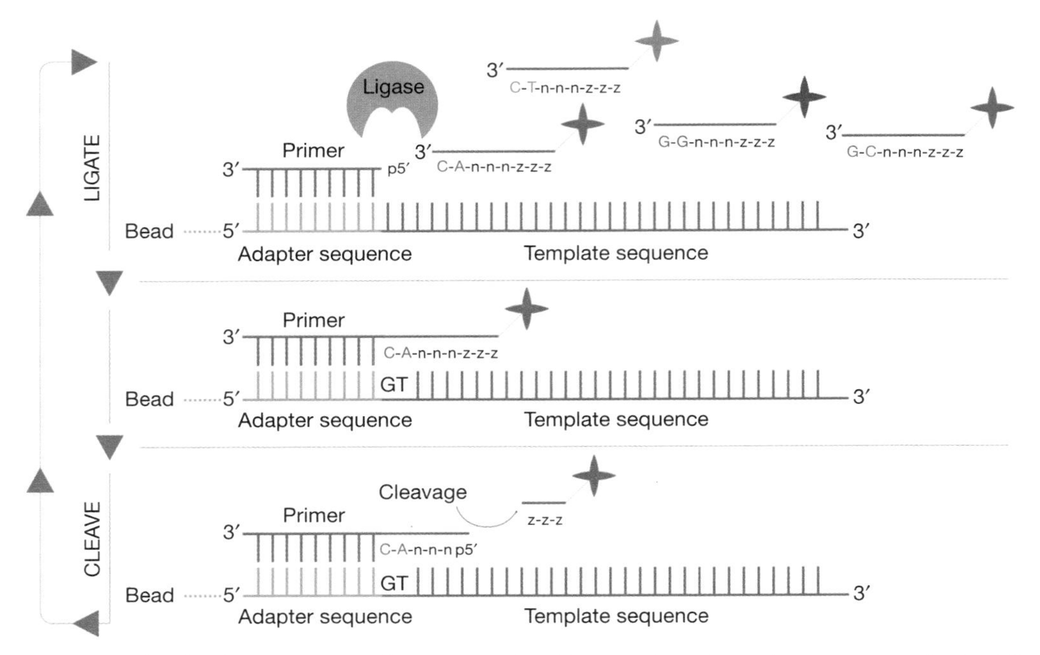

FIGURE 8. Oligo ligation detects specific 2-base nucleotide extensions from adapter sequence in the ABI SOLiD system.

as "z-z-z" in Fig. 8), allowing for ligation of another set of labeled oligos, which specifically identify bases 6 and 7. This can be repeated for several cycles, adding five bases each cycle and detecting the 2-base oligo that hybridized to each template.

Once the desired read length is achieved, the entire set of oligos is stripped off (a process that ABI calls a "reset"), and a new sequencing primer ($n - 1$) is applied that binds to the adapter at a position 1 base 5′ of the original sequencing primer. A new set of oligos are hybridized to the ($n - 1$) primer and detected. This is repeated for multiple cycles of terminator removal, oligo hybridization, and ligation, providing overlapping data for each position. Now position 1, 6, 11, 16, ..., have been interrogated with two different overlapping 2-base oligos. After the second set of ($n - 1$) oligos is hybridized, ligated, and detected, it is also stripped away and the entire process is repeated with three more sequencing primers ($n - 2$, $n - 3$, and $n - 4$) (see Fig. 9). This produces two oligo detections for every base in the template sequence.

The fluorescent labels on the oligos are designed in such a way that any combination of two overlapping oligos is uniquely defined, so that the base detected by the two oligos can be called unambiguously. However, the specific oligo detected at each position is not known, only its color (which represents any one of four different oligos). This 2-base "color space" encoding scheme has some interesting informatics consequences. To determine the entire sequence of a read, one base must be known. This first base is known from the primer sequence attached immediately upstream of the

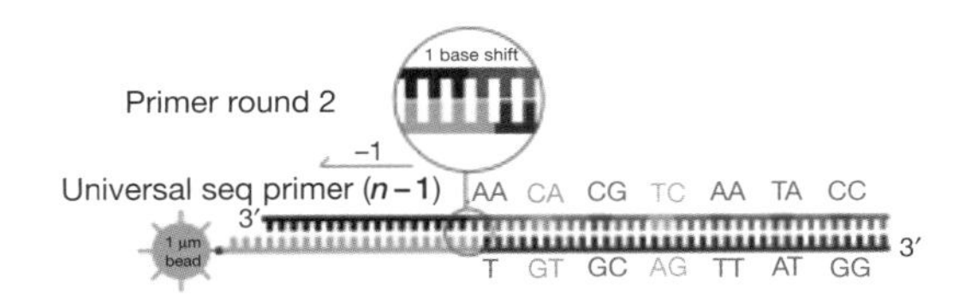

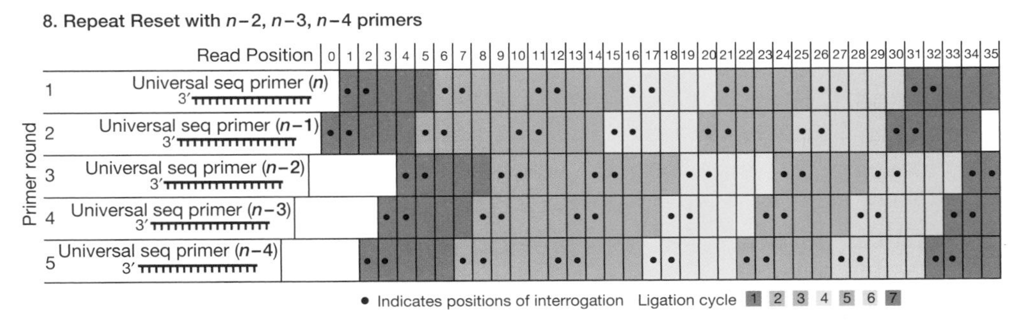

FIGURE 9. Multiple sets of SOLiD primer ligation events are combined to detect specific color space encoding of DNA sequence. (Reprinted, with permission, from ABI 2008.)

template fragment. If an error of fluorescent detection occurs, it will produce a single change in the oligo color data as compared with an aligned **reference sequence**, but a true mutation will produce two changes in the detected oligos (see Fig. 10). Therefore, the 2-base encoding system has error-correcting properties and is very well suited for the detection of single base mutations (SNPs). ABI claims system base-calling accuracy of 99.94%. ABI has produced software that stores SOLiD data as a series of oligo

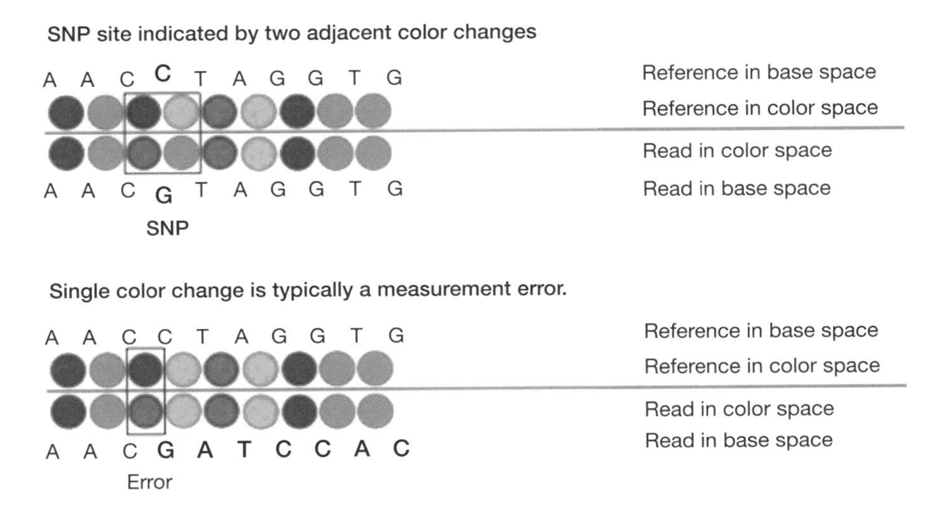

FIGURE 10. The SOLiD system accurately detects a SNP as two consecutive changes in color space data. (Reprinted, with permission, from ABI 2008.)

color calls rather than as DNA base calls. This "color space" data can be used to detect mutations such as insertions, deletions, and multibase nucleotide substitutions with greater specificity than standard DNA sequence alignment methods (ABI 2010).

Ion Torrent

The Ion Torrent Personal Genome Machine (PGM) was a latecomer to the NGS marketplace, introduced in December of 2010, but it achieved a sizeable market share very quickly because of several unique characteristics of the technology (and aggressive marketing by parent company Life Technologies). First, the Ion Torrent PGM was offered at a very inexpensive price compared with the other NGS platforms: The PGM is priced at <10% of the cost of an Illumina HiSeq, and the reagents consumed in a single run cost only about $500 (<5% of an Illumina HiSeq run). This puts it within the budgetary scope of ordinary laboratory equipment purchased by a single scientist, such as a centrifuge or a PCR machine, rather than an institutional-scale purchase. Second is speed. The PGM can produce data from a run in just 2 h (after an 8-h sample-preparation process).

The PGM uses beads to isolate and amplify single template molecules. Sample preparation for the PGM uses an emulsion PCR process, similar to the 454 machine. The Ion Torrent technology sequences large numbers of clonally amplified single molecules as templates on a solid surface (a picotiter plate), but it differs from 454, Illumina, and the ABI SOLiD in that it does not use fluorescent chemistry to detect individual bases. Instead, the PGM detects each nucleotide incorporation event by using a tiny semiconductor pH sensor, embedded in each well of the picotiter plate, to directly detect the H^+ ion released by the formation of the phosphodiester bond between each new base and the previous base of the growing DNA chain: hence the name "Ion Torrent." Each of the four DNA bases is added sequentially to the growing templates. Without fluorescent chemistry, the PGM uses only inexpensive natural nucleotides as reagents, and it does not need high-resolution cameras and complex image-processing software. Each base of sequence is captured directly as digital data. The PGM can increase its data throughput by packing more template molecules onto the surface of its sequencing chip with smaller picotiter wells and denser pH sensing circuitry. Ion Torrent has proposed that it will increase the base yield of the PGM by 100× per year. The PGM was released in December 2010 with the 314 chip with an output specified as 10 Mb (100,000 reads of 100-bp length). The 316 chip was released in April 2011 with an output specification of 100 Mb (1 million reads of 100-bp length). The Ion 318 chip (released in 2012) produces 200-bp reads, with a total output of 1 Gb of data. A larger machine, the Ion Proton, was also commercially released in 2012. It uses the same chemistry and the same read lengths as the PGM, but produces 10 times more reads per run.

Demonstration data released by Ion Torrent show 99% accuracy of aligned 100-bp reads. However, like the 454, the PGM sequences runs of identical bases in a single cycle, thus it is subject to insertion/deletion errors in homopolymer runs of eight or more bases. Also like the 454, the reads produced by a run on the PGM will vary in length depending on the number of repeated bases in the template sequence. The demonstration data show extremely even coverage of the *E. coli* genome, unaffected by variation in %G + C content.

The PGM was used to rapidly sequence toxic *E. coli* strains in the European food poisoning outbreak in May–June of 2011 (*Lancet*, 11 June 2011). Ten Ion 314 chips were run in 2 days to produce 18× coverage of the *E. coli* genome. Assembly with a mixture of reference-based and de novo methods produced a total of 364 contigs with N_{50} of 181,540 and a total contig (genome) length of 5 Mb (Life Technologies 2011). This genomic sequence allowed investigators to identify a novel aggressive, virulent, and antibiotic resistant *E. coli* strain containing a combination of genomes from a rare African strain called 0104:H4 known to cause serious diarrhea and another Shiga-toxin-producing strain (BGI 2011). The genomic sequence allowed for the design of a rapid and sensitive qPCR assay for the virulent strain. The short sample preparation and extremely fast run time of the PGM are advantageous in emergency situations, such as the outbreak of a new pathogen, but it also allows NGS to be a "hunch-driven" investigative tool, rather than a major undertaking that involves weeks of planning and consulting meetings. The PGM is also attractive to laboratories involved in routine amplicon sequencing, where access to rapid inexpensive sequencing is more useful than the very large amounts of data produced by other NGS platforms.

Paired-End and Mate-Pair Sequencing

The short reads from many of the NGS platforms create a disadvantage for some sequencing applications, especially de novo assembly of new genomes. Large repeats, segmental duplications, and regions of low-complexity sequence (which are all common in eukaryotic genomes) make it very difficult to assemble contigs that cover the entire genome from short reads. Even for small genomes, very deep coverage is required to create a single assembled contig for each chromosome. A **paired-end sequencing** strategy can greatly improve the efficiency of de novo genome assembly (see Chapter 6). The standard NGS sequencing method sequences one end of a short DNA fragment (fragments are typically 150–300 bp for Illumia and SOLiD systems). However, sequencing adapters are attached to both ends of each fragment. Using some clever chemistry, each vendor has developed a method to generate a separate sequence read from the opposite end of each fragment and to mark the pair of reads as coming from the same fragment. Using bioinformatics methods, it is possible to estimate the fragment size and therefore a distance between the locations of the two reads on the original genome. This distance information can be used to link contigs that do not overlap and to find unique genomic positions when one end of a DNA fragment contains a repeated sequence but the read from the other end contains unique sequence.

Paired-end reads can also be used to map insertion, deletion, and translocation events (structural variants) in mutated genomes. If the two reads from a pair map to positions on the reference genome significantly different from the expected fragment length, then a structural variant is likely.

A mate pair is similar in concept to a paired end, but is constructed by a more complex process. Instead of shearing genomic DNA into short 200–300-bp DNA fragments, much larger fragments are created, ranging from 2 kb up to 10 kb. Biotin-labeled adapters are added to the ends of these large fragments, and then the fragments are circularized by ligating the ends of the adapters together. The circles of DNA are then randomly sheared to create a set of 200–300-bp fragments, and the fragments that contain the biotin-labeled adapter are recovered. These fragments contain the left and right ends of the original large DNA fragment with the adapter in the middle. New sequencing adapters are added to both ends, and the fragment is sequenced using the paired-end protocol as described above. Interestingly, the mate-pair sequence reads the map back to the original genome in the reverse orientation as **paired-end sequences** (outward rather than inward facing) (see Fig. 11). **Mate-pair**

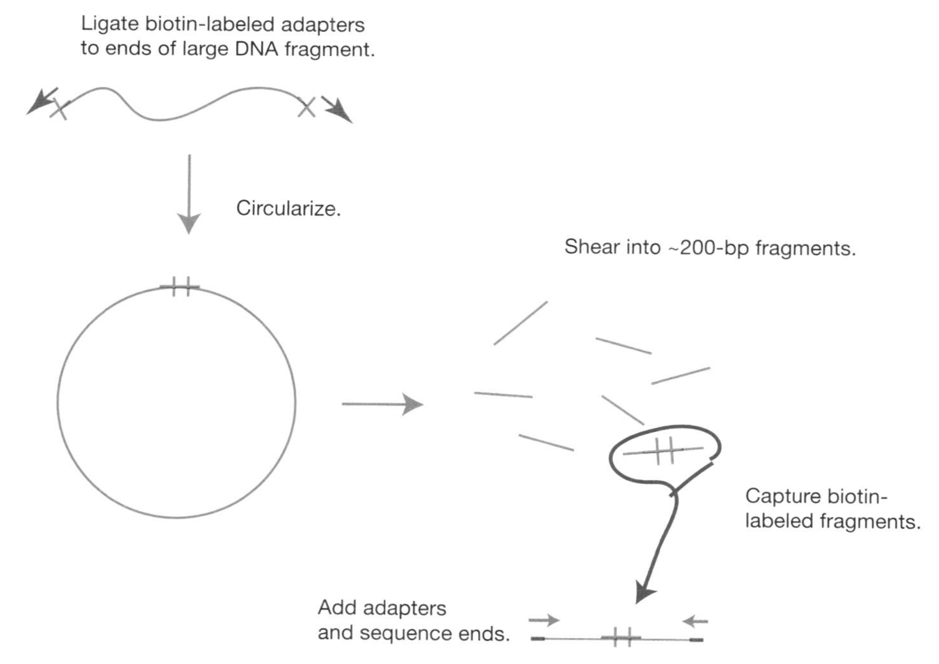

FIGURE 11. Mate-pair sequencing captures the ends of larger DNA fragments by circularizing then shearing the circles.

sequences can be used to link across larger gaps in de novo assemblies and deal with larger repeats in many different sequencing applications.

EXPERIMENTAL APPLICATIONS

Sequencing is a flexible data collection tool, analogous to microscopy, that can be used for an extremely diverse range of scientific applications by changing the type of sample and the manner of its preparation. Traditional Sanger sequencing allowed researchers to ask questions directly regarding DNA sequences, such as: "What proteins are encoded by this fragment of DNA?" or "What are the phylogenetic relationships among this group of organisms based on the sequences of this conserved gene?" Automated fluorescent sequencing allowed some additional questions to be addressed at the cost of substantial time and effort, such as: "What is the entire gene content and organization of the genome of this organism?" and "What genes are being expressed by this organism (or tissue)?"

NGS changed the scope of sequencing so dramatically that entirely new realms of experimental questions can be asked. Not only can new genomes be sequenced, but genome sequences from many different individuals of the same species can be compared in order to identify all variants (see Chapter 8). The genomes of tumors (or precancerous lesions) can be sequenced and compared with the normal germline cells of the same patient. NGS has become the data collection tool of choice for many different assays, replacing other technologies. NGS can be used on RNA, instead of **microarrays**, to measure the entire gene expression profile of any sample of cells (**RNA-seq**), with the additional value of discovering previously unannotated genes and quantitative measurement of alternative transcripts (see Chapter 10). NGS can be used, instead of genome tiling arrays, to provide a more accurate and higher-resolution readout for chromatin immunoprecipitation studies (**ChIP-seq**; see Chapter 9). NGS can also be used to study epigenetic changes in DNA methylation and modifications of **histone** proteins. The entire DNA content of environmental and medical samples can be sequenced in order to identify all microorganisms present as well as their complete genetic composition (metagenomics; see Chapter 11). NGS is an extremely dynamic field, with new applications constantly being created.

REFERENCES

Anderson S, de Bruijn MH, Coulson AR, Eperon IC, Sanger F, Young IG. 1982. Complete sequence of bovine mitochondrial DNA. Conserved features of the mammalian mitochondrial genome. *J Mol Biol* **156:** 683–717.

Applied Biosystems Incorporated (ABI). 2008. Principles of di-base sequencing and the advantages of color space analysis in the SOLiD system. Available at: http://seqinformatics.com.

Applied Biosystems Incorporated (ABI). 2010. A theoretical understanding of 2 base color codes and its application to annotation, error detection, and error correction. Available at: http://www3.appliedbiosystems.com/cms/groups/mcb_marketing/documents/generaldocuments/cms_058265.pdf (last accessed April 15, 2011).

Beijing Genomics Institute (BGI). 2011. BGI sequences genome of the deadly *E. coli* in Germany and reveals new super-toxic strain. BGI, Shenzhen, China, June 2, 2011. Available at: http://www.eurekalert.org/pub_releases/2011-06/bgia-bsg060211.php.

Bentley DR, Balasubramanian S, Swerdlow HP, Smith GP, Milton J, Brown CG, Hall KP, Evers DJ, Barnes CL, Smith AJ, et al. 2008. Accurate whole human genome sequencing using reversible terminator chemistry. *Nature* **456:** 53–59.

Cox-Foster DL, Conlan S, Holmes EC, Palacios G, Evans JD, Moran NA, Quan PL, Briese T, Hornig M, Lipkin WI, et al. 2007. A metagenomic survey of microbes in honey bee colony collapse disorder. *Science* **318:** 283–287.

Furey WS, Joyce CM, Osborne MA, Klenerman D, Peliska JA, Balasubramanian S. 1998. Use of fluorescence resonance energy transfer to investigate the conformation of DNA substrates bound to the Klenow fragment. *Biochemistry* **37:** 2979–2990.

Gilbert W, Maxam A. 1973. The nucleotide sequence of the *lac* operator. *Proc Natl Acad Sci* **70:** 3581–3584.

Gilbert MT, Tomsho LP, Rendulic S, Packard M, Drautz DI, Sher A, Tikhonov A, Dalén L, Kuznetsova T, Schuster SC, et al. 2007. Whole-genome shotgun sequencing of mitochondria from ancient hair shafts. *Science* **317:** 1927–1930.

Gladwell M. 2008. Nathan Myhrvold and collective genius in science. *The New Yorker* (12 May).

Goldstein IF, Goldstein M. 1978. *How we know: An exploration of the scientific process.* Plenum Press, New York.

Green RE, Krause J, Ptak SE, Briggs AW, Ronan MT, Simons JF, Du L, Egholm M, Rothberg JM, Paunovic M, Pääbo S. 2006. Analysis of one million base pairs of Neanderthal DNA. *Nature* **444:** 330–336.

Henikoff S. 1984. Unidirectional digestion with exonuclease III creates targeted breakpoints for DNA sequencing. *Gene* **28:** 351–359.

Hyde R. 2011. Germany reels in the wake of *E. coli* outbreak. *Lancet* **377:** 1991.

Kuhn TS. 1996. *The structure of scientific revolutions*, 3rd ed. University of Chicago Press, Chicago.

Ley TJ, Mardis ER, Ding L, Fulton B, McLellan MD, Chen K, Dooling D, Dunford-Shore BH, McGrath S, Wilson RK, et al. 2008. DNA sequencing of a cytogenetically normal acute myeloid leukaemia genome. *Nature* **456:** 66–72.

Life Technologies. 2011. Shiga toxin–producing *Escherichia coli*. Application Note. Life Technologies, Carlsbad, CA. Available at: http://www.iontorrent.com/lib/images/PDFs/co23298_e coli.pdf (last accessed April 15, 2011).

Malde K. 2011. Flower: Extracting information from pyrosequencing data. *Bioinformatics* **27:** 1041–1042.

Margulies M, Egholm M, Altman WE, Attiya S, Bader JS, Bemben LA, Berka J, Braverman MS, Chen YJ, Rothberg JM, et al. 2005. Genome sequencing in microfabricated high-density picolitre reactors. *Nature* **437:** 376–380.

Maxam AM, Gilbert W. 1977. A new method for sequencing DNA. *Proc Natl Acad Sci* **74:** 560–564.

Miller MJ, Powell JI. 1994. A quantitative comparison of DNA sequence assembly programs. *J Comput Biol* **1:** 257–269.

Min Jou W, Haegeman G, Ysebaert M, Fiers W. 1972. Nucleotide sequence of the gene coding for the bacteriophage MS2 coat protein. *Nature* **237:** 82–88.

Nagalakshmi U, Wang Z, Waern K, Shou C, Raha D, Gerstein M, Snyder M. 2008. The transcriptional landscape of the yeast genome defined by RNA sequencing. *Science* **320:** 1344–1349.

Nossa CW, Oberdorf WE, Yang L, Aas JA, Paster BJ, Desantis TZ, Brodie EL, Malamud D, Poles MA, Pei Z. 2010. Design of 16S rRNA gene primers for 454 pyrosequencing of the human foregut microbiome. *World J Gastroenterol* **16:** 4135–4144.

Nyrén P. 2007. The history of pyrosequencing. *Methods Mol Biol* **373:** 1–14.

Pollack A. 2003. Company says it mapped genes of virus in one day. *NY Times* (August 22).

Ronaghi M, Karamohamed S, Pettersson B, Uhlen M, Nyrén P. 1996. Real-time DNA sequencing using detection of pyrophosphate release. *Anal Biochem* **242:** 84–89.

Sanger F, Coulson AR. 1975. A rapid method for determining sequences in DNA by primed synthesis with DNA polymerase. *J Mol Biol* **94:** 441–448.

Sanger F, Nicklen S, Coulson AR. 1977. DNA sequencing with chain-terminating inhibitors. *Proc Natl Acad Sci* **74:** 5463–5467.

Smith LM, Sanders JZ, Kaiser RJ, Hughes P, Dodd C, Connell CR, Heiner C, Kent SB, Hood LE. 1986. Fluorescence detection in automated DNA sequence analysis. *Nature* **321:** 674–679.

Thomas RK, Greulich H, Yuza Y, Lee JC, Tengs T, Feng W, Chen TH, Nickerson E, Simons J, Egholm M, et al. 2005. Detection of oncogenic mutations in the *EGFR* gene in lung adenocarcinoma with differential sensitivity to EGFR tyrosine kinase inhibitors. *Cold Spring Harb Symp Quant Biol* **70:** 73–81.

Wang C, Mitsuya Y, Gharizadeh B, Ronaghi M, Shafer RW. 2007. Characterization of mutation spectra with ultra-deep pyrosequencing: Application to HIV-1 drug resistance. *Genome Res* **17:** 1195–1201.

Wang J, Wang W, Li R, Li Y, Tian G, Goodman L, Fan W, Zhang J, Li J, Wang J, et al. 2008. The diploid genome sequence of an Asian individual. *Nature* **456:** 60–65.

Wheeler DA, Srinivasan M, Egholm M, Shen Y, Chen L, McGuire A, He W, Chen YJ, Makhijani V, Rothberg JM, et al. 2008. The complete genome of an individual by massively parallel DNA sequencing. *Nature* **452:** 872–876.

Wu R, Taylor E. 1971. Nucleotide sequence analysis of DNA. *J Mol Biol* **57:** 491–511.

WWW RESOURCES

http://hackage.haskell.org/package/flower The flower package home page. Flower (FLOWgram ExtractoR tools) reads files in SFF-format.

http://www.dnabaser.com/download/SFF%20tools/index.html SFF Workbench (initial name "454 SFF Tools") is an easy to use SFF file viewer, editor, and converter.

http://www.ncbi.nlm.nih.gov/Traces/trace.cgi?cmd=show&f=formats&m=doc&s=format#sff Trace Archive, v4.2. National Center for Biotechnology Information (NCBI), U.S. National Library of Medicine.

2

History of Sequencing Informatics

Stuart M. Brown

EARLY INFORMATICS METHODS

Biochemical methods to determine the amino acid sequence of proteins were developed much earlier than DNA sequencing methods. Per Edman developed a simple chemical method (Edman degradation) to remove and identify amino acids one by one from the amino-terminal end of a protein (Edman 1950). Frederick Sanger used a method of partial hydrolysis and amino-end labeling to determine the complete sequence of the two polypeptides in the insulin protein (Sanger and Tuppy 1951a,b).

By the mid 1960s, a substantial number of protein sequences were known. Margaret Dayhoff gathered all known protein sequences into the first edition of her *Atlas of Protein Sequence and Structure* (Eck and Dayhoff 1966). Walter Fitch developed a computer program to search for "evolutionary homology" between protein sequences by comparing amino acid strings (words) of variable length. "The computer compares all possible consecutive amino acid sequences from one protein with all possible such sequences from the other protein." Fitch's program used a scoring system that counted each amino acid mismatch as the minimum number of mutations that could convert the codon for one into that for the other. It should be noted that this work followed very shortly after the final determinations of the DNA codons for each of the amino acids by Severo Ochoa, Philip Leder, Marshall Nirenberg, and H. Gobind Khorana (Nirenberg et al. 1965). The Fitch program did not allow for gaps within the words being compared, and he recommended a word size of 30 amino acids as the default. Fitch also developed a program to identify insertion/deletion events in the **alignment** of closely related proteins (Fitch 1966c).

Needleman and Wunsch (1970) presented a very important paper on computing pairwise amino acid similarities. They outlined a method for comparing two amino acid sequences in a two-dimensional (2D) matrix by computing all possible frames of alignment with all possible gaps. This is a dynamic programming approach to the problem of **sequence alignment**. However, their method, as described in the original 1970 paper, does not fully implement an appropriate scoring scheme to penalize gaps and amino acid mismatches (nonidentical pairs). This early Needleman–Wunsch method also lacks a built-in significance test for alignments, and it does not normalize scores for the lengths of amino acids being aligned. Sankoff (1972) improved the Needleman–Wunsch method by creating a gap penalty (deletion/insertion index) that can be minimized or set to a threshold value, as well as improving the computational efficiency of the **algorithm**.

Smith and Waterman (1981) developed a greatly improved method of pairwise alignment for sequences using a dynamic programming approach similar to that of Needleman–Wunsch, but only calculating positive alignment scores for regions of the matrix. This allows for short segments within longer sequences to have positive scores in a local alignment ("maximally homologous subsequences"), with full consideration of gaps (insertions and deletions of any length) and adjustable gap penalties. The **Smith–Waterman alignment** remains the best, mathematically rigorous method to find the optimal gapped local alignment for any two segments of DNA (or protein) sequence.

DATABASE SEARCHING

The Smith–Waterman algorithm finds the optimal alignment between two sequences, but it is computationally demanding because it searches all possible frames of alignment with all possible gaps. As collections of DNA and protein sequences grew larger in the 1980s, many scientists needed to routinely compare sequences against large databases. The Smith–Waterman method proved to be too slow for routine searches against tens of thousands of sequences. Pearson and Lipman created the FASTP sequence alignment program in 1985 (Lipman and Pearson 1985) to compare protein sequences with the National Biomedical Research Foundation (NBRF) protein sequence database. FASTP performed searches two orders of magnitude faster than Smith–Waterman. In 1988, Pearson and Lipman published a new program, called FASTA (Pearson and Lipman 1988), which allowed for the searching of both protein and DNA databases. It could also compare a protein query sequence with a DNA database by automatically translating the DNA database (in all reading frames) as it was searched. The FASTA program was freely released as C source code and as precompiled executables for UNIX, VAX/VMS, and IBM DOS operating systems. Versions of FASTA that run on the IBM PC were limited to query sequences with a maximum of 2000 bases or amino acids. The investigators report that FASTA was capable of searching the entire **GenBank** DNA database (Release 48) with a 660-nucleotide query sequence on an IBM PC AT computer in <15 min.

The FASTA program makes use of a lookup table of short words with a word size, known as ***ktup***, that is created from each pair of sequences being compared.

Alignments are only created at positions sharing identical words of *k*tup length. Then these alignments are extended using additional identical letters, mismatch letters followed by more matching letters, or conservative amino acid changes in protein sequences (calculated using the PAM250 scoring matrix). If a pair of sequences contains more than one word match region, FASTA tries to join the aligned regions with a gap penalty. Once aligned regions have been extended and joined as much as possible, FASTA calculates an overall alignment score, which is used to rank all sequences in the database versus the query sequence. The highest-scoring database sequences are then realigned to the query using a modified version of the Smith–Waterman method that only considers alignments in a band that contains the initial high-scoring alignment region.

The FASTA package also provides a program (RDF2) to estimate the statistical significance of alignments. This is extremely important because heuristic alignment methods such as FASTA may lose specificity as they increase sensitivity. A small *k*tup value may allow FASTA to locate sequences in a database that match a query, but a statistical standard is needed to evaluate if these matches are biologically meaningful. The statistical test used by RDF2 is based on the concept of making random permutations of the letters in one sequence and repeating the FASTA alignment. Under most conditions, a query sequence with a good alignment score to a database sequence will no longer show high-scoring alignment when the database sequence is randomly shuffled. If the alignment score is not significantly reduced by shuffling, then the alignment is likely due to biased composition of the two sequences (i.e., A + T- or G + C-rich regions, or simple sequence repeats) and not the result of functional and evolutionary similarity between the sequences.

During the 1980s and 1990s, the growth of DNA sequence databases, such as GenBank, outpaced the speed improvements of computer CPUs; consequently, the amount of time required for a database search with the FASTA program increased. Investigators also needed to make database searches in large batches to annotate large amounts of sequence data, such as genome sequencing projects and mRNA sequencing (expressed sequence tags). A faster DNA similarity search algorithm was needed. In response to this need for faster similarity searches, Stephen Altschul and coworkers at the National Center for Biotechnology Information (NCBI, a division of the NIH and the home of GenBank) developed the Basic Local Alignment Search Tool (BLAST) algorithm (Altschul et al. 1990). In addition to greater speed, BLAST provides improved statistical estimates of similarity scores under an appropriate random sequence model (Karlin and Altschul 1990).

BLAST is a local alignment method, similar in its basic concept to Smith–Waterman and FASTA. The BLAST method uses an estimate of the size of aligned sequence segments that are due to random matches between a particular query sequence and a database. BLAST ignores short local alignments that do not have an alignment score greater than this random threshold. BLAST gains additional speed from heuristic methods to locate regions where aligned segments may occur. Both the query sequence and the database are broken up into short words of length *w*. Then each sequence in the database is quickly scanned for words that match any words in the query sequence. Unlike FASTA, which only finds matches between identical words, the BLAST word-matching step uses a scoring system that allows for some mismatches as long as the overall alignment score between words is above an empirically determined threshold. Longer word pairs can have more mismatches. Matches between protein sequences use an amino acid similarity matrix such as the PAM120 matrix (Dayhoff et al. 1978), whereas the bases of DNA alignments are scored as a simple identity (+5) or mismatch (−4). Rather than compute scores between each word from the query sequence and each database word, BLAST pre-computes a list of all possible words that can match each word in the query sequence with an above-threshold score. This extended word list derived from the query sequence can then be compared with the list of words in each database sequence in a single-step hash table lookup, which is computationally very fast (as long as adequate memory is available).

When a high-scoring word pair is found between the query and a database sequence, it is extended in both directions to form an ungapped alignment until additional pairs of residues do not increase the overall score; thus the matched word pair functions as a seed for the more computationally expensive pairwise alignment process. This aligned region is known as a maximal segment pair (MSP). The alignment score of the MSP is compared with the threshold for random matches, so that only significant MSPs are retained. Once all of the sequences in the database have been scanned for word matches and all of the word pairs have been extended to MSPs, then alignments are ranked by MSP alignment scores.

BLAST is most powerful when searching for protein sequence similarity, but it also offers great speed in DNA–DNA searching with acceptable sensitivity and specificity. DNA words are 11 bases long by default ($w = 11$), but greater computational acceleration is achieved by packing 4 nucleotides into a single byte and searching the database bytewise for 2-byte matches (8 bases), then rescanning these matches for full *w*-mers. A great improvement in computational speed can also be achieved by storing all of the words for a single database in memory and repeating the search with many different query sequences.

The raw scores produced by BLAST represent the local alignment values of segment pairs between the query and database sequences. These values are only meaningful when compared with a statistical model of random sequence alignments. In other words: Is this match between query and database sequence better than the match between two unrelated sequences? Is it better than the expectation for a random query sequence to match something in the database? The alignment scores of MSPs from a BLAST search follow an extreme value distribution (Gumbel 1958; http://www.ncbi.nlm.nih.gov/BLAST/tutorial/Altschul-1.html). From this distribution, an e-value can be calculated that represents the likelihood that a given match score would occur by chance in a search with a random (unrelated) query sequence against this database:

$$E = Kmne^{-\lambda S}.$$

In this equation, K represents the size of the database and λ represents the alignment scoring system, m is the length of the query sequence, n is the length of the aligned database sequence, and S is the alignment score. The smaller the e-value, the less likely it is that a score could occur by chance. For an alignment with an e-value of 1, a random query sequence of a similar length, searched against this database, would be expected to produce at least one alignment with an equivalent score. Biologists often set an e-value cutoff of 0.05 as a significance threshold for an alignment, but the biological significance of an e-value score is extremely dependent on the context of the search. A BLAST search must be approached like any experiment or statistical test, with a hypothesis to be tested and a predetermined value for the test result, which will disprove the null hypothesis.

BLAST has proven to be an extremely useful tool for investigating many different types of experimental questions. It is used to search for matches between both extremely similar sequences, such as matching **sequence fragments** back to the known genome sequence of a species; as well as elucidating evolutionary relationships between distantly related sequences, which may be from different species or may be members of a gene family within a single genome.

BLAST is a heuristic method, thus it does not guarantee the discovery of all optimal alignments between a query sequence and a database. The values for word size, minimum word matching score, and minimum MSP alignment score can be set empirically, but there is a trade-off between sensitivity and specificity. In practice, biologists are willing to accept a moderate level of false-positive matches (lower specificity) in exchange for a low chance of missing biologically meaningful matches (higher sensitivity). Databases of biological sequences are highly nonrandom, with locally biased base composition and repeated sequence elements, which tend to create false-positive matches. To compensate for this problem, BLAST automatically filters out all 8-base words from the query sequence that occur in the database at elevated frequency. BLAST can use a secondary database of known repeated elements (and vector sequences) to prescreen query sequences. Regions of the query that match the repeat database are filtered out of the main database search, but the repeat match is reported in the final output.

BLAST was developed by a team at the NCBI, and in addition to freely releasing the C language source code, they created a free web-based service where anyone can run a BLAST search against a selection of databases hosted at the NCBI including

GenBank DNA sequences, translated GenBank protein-coding sequences, EST sequences, whole genomes for specific organisms, and data sets limited to specific taxonomic groups. Since its inception, the NCBI BLAST has consistently been the most used piece of bioinformatics software in the world.

The speed of BLAST database searches is proportional to the product of the length of the query sequence and the total length of all of the sequences in the database being searched. Because sequence databases increase in size more rapidly than computer processors increase in speed, the computing effort to run each BLAST search increases over time. In 1997, the NCBI team created a new version of BLAST, known as Gapped BLAST (Altschul et al. 1997), which improved the detection of gapped alignments and dramatically improved the speed of BLAST, while also increasing the sensitivity for similarities between distantly related sequences. The essence of the gapped BLAST innovation was to require two nearby, nonoverlapping word matches as a seed for creating an alignment between query sequence and database sequence. This greatly reduces the number of word matches that must be extended and tested to see if they produce significant alignments, the slowest part of the BLAST method (using >90% of execution time). This method naturally allows for gaps between the two initial matching words; in addition, high-scoring MSPs are extended by a pairwise alignment method. The maximum distance between the two initial matching words is an empirical parameter.

THE STADEN PACKAGE

The development of computational tools to handle DNA sequence data closely parallels the development of the technology to determine the sequence of DNA. The earliest programs designed to work with DNA sequences were developed by Rodger Staden, working at the MRC Laboratory of Molecular Biology (also the home of Sanger and of Watson and Crick), to facilitate the work on genes and small genomes being sequenced by application of Sanger or Maxam–Gilbert sequencing (Staden 1977). Staden's very first package of sequence informatics software (written in PDP FORTRAN) had the following capabilities: (1) storage and editing of a sequence, (2) producing copies or portions of the sequence in single- or double-stranded form, (3) translation of DNA into amino acid sequence, (4) searching the sequence for any particular shorter sequences (e.g., restriction enzyme sites, (5) analysis of codon usage and base composition, (6) comparison of two sequences for homology, (7) locating regions of sequences that are complementary, and (8) translation of two aligned sequences with the printout showing amino acid similarities.

The SEQFIT program in the 1977 Staden software package may be the first DNA alignment program ever written. As such, it deserves a bit of special attention in this book. This program used an exhaustive search strategy—the "slide and count" method—to find all possible frames of overlap between two sequences and counted the matched bases (Staden 1977):

> The operator defines the [target] region he wishes to compare with the [test sequence] string and specifies the minimum degree of similarity required, expressed as a percentage. The program places the string alongside the defined region in every possible position and counts the total number of identical characters in adjacent positions. If this total, or score, expressed as a percentage of the length of the string, is greater than or equal to the percentage required, the program remembers the position at which it occurred.

This program was also capable of automatically comparing the complement of the test sequence with the target. It is interesting to note that the original Staden DNA alignment program did not take advantage of existing bioinformatics methods developed to align protein sequences such as the dynamic programming approach used by the Needleman–Wunsch method.

The basic strategy of sequencing large regions of DNA by finding overlapping regions between sequence fragments and assembling them into a longer contiguous **consensus sequence** (**contig**) was already well established by 1979 (Staden 1979). The 1982 version of the Staden software creates a database of original sequences (reads) as well as creating joined and edited contig sequences. This idea of preserving original "experimental" data files and creating new files to contain edits, contigs, and consensus sequences has been maintained in virtually all subsequent software for managing DNA sequencing data.

Rodger Staden and coworkers updated their software package in 1995 and renamed it the Genome Assembly Program (GAP) (Bonfield et al. 1995). Many new features were added including a tool for assembly of **shotgun sequence reads**, a tool for directed assembly, and a tool to "find internal joins," which finds regions of possible overlaps between contigs. However, the shotgun assembly tool is only able to add individual reads one by one to contigs already assembled in the internal database; it does not perform de novo all against all assembly of individual reads (global assembly). Contigs are only joined by the shotgun algorithm if a single added read aligns well to two existing contigs and forms a bridge between them. A tool called "find read pairs" allows sequence reads that are taken from the opposite ends of a cloned DNA fragment to create a join between two contigs (to form a scaffold) even when there is a gap in the overlapping sequences. The program also included a guided assembly algorithm, which detected unfinished or problematic areas of a **sequence assembly** and suggested experimental strategies (additional sequencing reactions with custom primers). An important addition was a graphical user interface with an interactive sequencing alignment/assembly editor. This program was designed to manage sequence reads for a sequencing project at the scale of a single cosmid, which is ~20–40 kb of total sequence, which might be covered by 200–1000 reads. The GAP package was written in ANSI C and FORTRAN77, and the user interface made use of the Tcl and Tk libraries, which emulate the

Motif X-windows graphical environment on Sun, DEC, and UNIX workstations. The code was made freely available by the Staden laboratory.

The GAP program included a very important novel feature, which used numerical estimates of base-calling accuracy to automatically build consensus sequences with the best quality data, rather than requiring manual editing of every incorrect base or using simple majority rule wherever disagreements occur. Without accuracy scores, assembly of overlapping reads for shotgun, primer walking, deletion subclone, or even restriction fragment sequencing strategies required time-consuming manual editing of every disagreement between bases in aligned reads. For manual sequencing, this step required laboratory staff to visually inspect autoradiograph films to try to identify gel-reading or transcription errors. For ABI automated sequencers, the chromatogram files for each read must be visually inspected and incorrect base calls flagged or corrected. The Staden GAP package allowed visual inspection of chromatogram files directly on the computer screen aligned with contig sequences with base-call discrepancies highlighted. "Although being able to show the original traces is necessary, having numerical estimates of base accuracy is the key to further automation of data handling for sequencing projects" (Bonifield and Staden 1995b). The GAP package contains a simple program, EBA (Estimate Base Accuracy), that computes the accuracy of each base call directly from the chromatogram data as the normalized ratio of the area under the highest peak (the called base) to the area under the second highest peak at the same position in the chromatogram. The investigators of the GAP program acknowledged ongoing work by other groups to develop more reliable estimates of base-calling accuracy (Lawrence and Solovyev 1994; Lipshutz et al. 1994), and intentionally designed GAP to be able to incorporate quality scores produced by other software. These quality scores can then be used both to resolve base differences in aligned reads when creating a consensus sequence as well as to build contigs when overlapping portions of sequence reads have some mismatching bases.

The GAP program was updated and renamed GAP4 in 1998 (Staden et al. 1998), which became one of the most popular tools for DNA sequencing informatics. GAP4 includes a graphical interface, scripts for standard data processing tasks such as vector removal and quality-based trimming of sequence reads, a database (and custom file format) for storing all data related to sequencing projects, and an integrated collection of useful software tools from other investigators. GAP4 contains methods for **de novo assembly** of all sequence reads into contigs, tools for checking and editing assemblies, and methods for the automated design of custom experiments to remedy gaps or poor-quality data in order to achieve a complete high-quality consensus sequence covering the entire target region of a sequencing project (see Fig. 1). The design of GAP4 as a master interface, which can easily integrate other software as supporting modules, allowed this program to be rapidly upgraded and customized by the development team and by users, so that it has remained popular for many

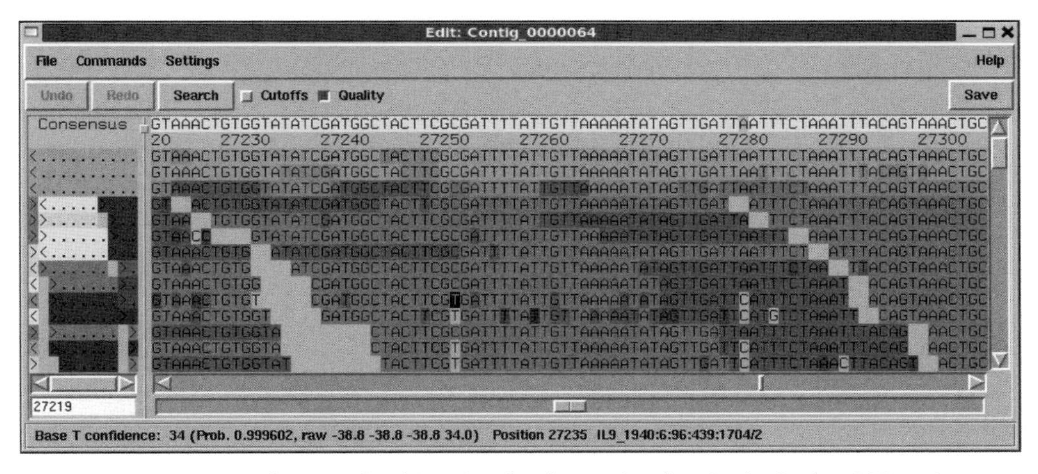

FIGURE 1. An X-Windows graphical interface for the contig editor in the Staden GAP package.

years. Important add-on modules used by GAP4 include assembly programs CAP2, CAP3, FAKII, and Phrap; identification and removal of vector sequences with cross_match; screening for contaminants with BLAST; screening for repeats with RepeatMasker; and quality scoring with Phred or ATQA.

A new version of the Staden sequencing informatics package, called GAP5, was released in 2010 with the capability to assemble and edit billions of **short reads** generated by **next-generation sequencing** machines (Bonfield and Whitwham 2010).

Intelligenetics and Wisconsin/GCG Packages

In the early 1980s, several different commercial software packages were developed to aid molecular biologists with DNA sequencing projects and the computational analysis of DNA sequences. At Stanford, the Intelligenetics company was created to develop and distribute a DNA sequence assembly system known as GEL (Clayton and Kedes 1982). The GEL system is similar in concept to the Staden software but automates many of the tasks involved in assembly of random fragments, including automatically testing a sequence fragment and its reverse complement for overlaps to all other sequence fragments and contigs in a project. Like the Staden system, original sequences, entered as experimental data, are stored in a database, while edits and contigs are stored in separate project files. Intelligenetics marketed a suite of programs including GEL, SEQ (restriction maps, translation, and sequence alignment), PEP (protein sequence analysis), MAXIMIZE (design an optimal DNA sequencing strategy from a restriction map), and GENESIS (laboratory data management).

The Genetics Computer Group (GCG) at the University of Wisconsin developed an integrated suite of molecular biology software written in FORTRAN for the VAX computer using the VMS operating system (Devereux et al. 1984). The GCG software system is noteworthy for several reasons. The package was designed in a modular fashion, with a large set of small, single-function programs (34 programs are described in the 1984 paper), all of which shared a set of common file formats and similar input/output interface conventions. The software was distributed on an explicitly paid basis: "A fee of $2,000 for non-profit institutions or $4,000 for industries is being charged for a tape and documentation for each computer on which UWGCG software is installed." Yet the concept of an open source community was also directly addressed: "UWGCG software is designed to be maintained and modified at sites other than the University of Wisconsin. The program manual is extensive and the source codes are organized to make modification convenient. Scientists using UWGCG software are encouraged to use existing programs as a framework for developing new ones. Our copyright can be removed from any program modified by more than 25% of our original effort." Thus a formal open source policy existed in bioinformatics software long before the term "open source" was formally adopted by the Open Source Initiative in 1998.

The Wisconsin Package included sequence editing programs, Seqed and Assemble, that were specifically designed to aid DNA sequencing projects with a screen editor to simplify manual data entry and error checking of DNA sequences and assembly of overlapping fragments. The GCG package was also explicitly aware of the file formats used by other bioinformatics software from Staden and Intelligenetics, as well as the GenBank and EMBL databases; and it contained programs to convert files to and from these formats. GCG file formats contained only human-readable text, compatible with both GenBank and EMBL, and made use of the (not yet standard) IUB-IUPAC nucleotide ambiguity symbols.

Desktop Sequence Assembly and Editing Software

The Staden, Intelligenetics, and GCG software packages provided useful toolkits for researchers involved in routine DNA sequencing projects. However, these programs were designed to run on large central computers using client-server operating systems, with text-based (or X-Windows) user interfaces via terminals. As desktop computers became common in the laboratory in the late 1980s, researchers desired simpler software with graphical user interfaces. The use of desktop computers for sequencing informatics was very popular among molecular biologists, not just because of possible savings in computer costs, but by allowing laboratory staff (faculty, postdoctoral fellows, graduate students, and technicians) to manage their own informatics needs without reliance on information technology professionals. In particular, the ABI-automated DNA sequencing machines used Macintosh computers to operate the machine control, image processing, and base-calling software; thus every laboratory with an ABI sequencer had a Macintosh computer.

Key features of software for DNA sequencing projects include finding and removing vector sequences at the ends of sequence reads, trimming low-quality sequences from the ends of reads, identifying overlapping reads and assembling contigs, and visually inspecting contigs to resolve base disagreements between different reads in aligned positions (sequence editing). With ABI-automated sequencers, it was also possible to inspect fluorescence data from chromatogram files directly on the computer, as opposed to reinspecting autoradiographic films to resolve errors in manual sequencing. Sequence editing software is used to identify and create **cloning** and sequencing strategies that will extend the sequence of contigs to fill gaps as well as to add additional sequence reads to cover low-quality regions of the assembly.

In 1991 the Gene Codes Corporation introduced the commercial program Sequencher for the analysis and assembly of DNA fragments. Sequencher was a Macintosh program that read ABI chromatogram files, automated the process of identifying and trimming vector and low-quality sequence from each sequence read, and had an innovative **fragment assembly** algorithm with adjustable parameters (minimum number of bases of overlap, minimum percent identity). The removal of vector sequence was based on alignment/overlap of the ends of sequence reads to a manually entered vector sequence at the cloning site. The removal of low-quality sequence from the ends of reads depends on the use of *N* letters in the sequence files produced by ABI base-calling software, where chromatogram data are ambiguous. The researcher can manually set window size and percent of N base calls to trim from the ends of sequences; for example, "Trim from the 3′ end until the last 25 bases contain less than 3 ambiguities."

Sequencher also allowed researchers to visualize chromatogram files in an assembly and manually edit base discrepancies in overlapping sequence fragments based on a visual assessment of sequence quality (see Fig. 2). Sequencher can also

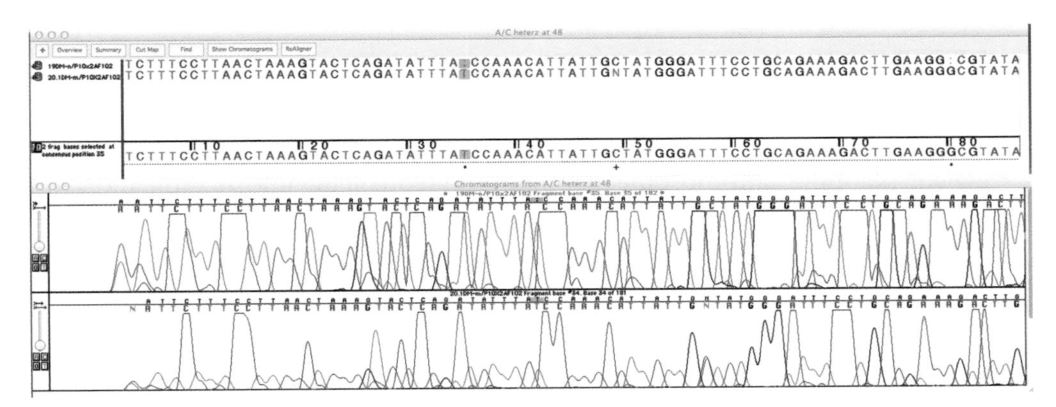

FIGURE 2. A graphical view on a Macintosh computer of the contig editor of the Sequencher program (Gene Codes) showing aligned chromatograms for two sequence reads.

build automatic consensus sequences by using the most common base in each position among a set of overlapping fragments. This "consensus by plurality" algorithm requires a high level of sequence **coverage** across the entire region of interest, which, in turn, requires many cloning and subcloning steps (and many lanes of sequencing data) in order to produce a high-quality sequence.

A review paper written in 1994 (Miller and Powell 1994) compared the performance of 10 different commercially marketed DNA sequence assembly programs, including five for the Macintosh Computer: Sequencher (Gene Codes, Inc., Ann Arbor, MI), MacVector (International Biotechnologies, New Haven, CT), GeneWorks (Intelligenetics, Mountain View, CA), Lasergene (DNAStar, Madison, WI), AutoAssembler (Applied Biosystems, Foster City, CA); and one for Windows PC (PC/Gene; Intelligenetics, Hilton Head Isle, SC). All of these programs allowed assembly of a few dozen to a few hundred unordered DNA sequence reads into contigs a few thousand bases long; the size of a typical single-gene sequencing project.

Phred/Phrap

The **Human Genome Project** (HGP) was a U.S. Government–led effort, officially launched in April 1990, to sequence the entire human genome. The primary goal of the project was "analyzing the structure of human DNA and determining the location of the estimated 100,000 [*sic*] human genes." The Project was jointly led by the NIH and the Department of Energy and funded at $200 million per year for 15 years (anticipating completion of the Project in 2005). One of the critical early milestones for the HGP was the development of informatics tools for the storage and analysis of large-scale sequence information. In 1990, the 3-billion-base human genome was larger than any existing set of sequence data, thus it was clear that new databases and analysis tools must be developed to work with data at this scale. The official informatics goals, as stated in the joint NIH/DOE 5 year plan (NIH 1990), are as follows:

- 5 YEAR GOAL: Develop effective software and database designs to support large-scale mapping and sequencing projects.
- Create database tools that provide easy access to up-to-date physical mapping, genetic mapping, chromosome mapping, and sequencing information and allow ready comparison of the data in these several data sets.
- Develop algorithms and analytical tools that can be used in the interpretation of genomic information.

With funding from the HGP, Phil Green, Brent Ewing, David Gordon, and others at the University of Washington, Seattle, working closely with the Genome Sequencing Center at Washington University, Saint Louis, developed a set of informatics tools for processing raw **Sanger sequences** collected by ABI sequencing machines and for assembling overlapping reads into larger sequence fragments (contigs). These tools, known as Phred, Phrap, cross_match, and consed, became the standard sequence processing and assembly software for all of the laboratories collaborating in the HGP. These tools were released as C source code (under a free academic license or paid commercial license) suitable for compilation by skilled users, typically on UNIX-based computers (including Linux, DEC Alpha, HP-UP, Irix, and Solaris). The Consed graphical alignment editor makes use of X11 and Motif graphics, which may be implemented directly on a UNIX workstation, an X-terminal, or via X11 communication between a desktop computer and a UNIX server.

The ABI sequencing machines use Sanger sequencing with four-color fluorescent dye terminators. Each labeled DNA fragment creates a peak of fluorescence in one of the four colors as it moves past set of detectors at the bottom of the sequencing gel (or capillary tube). These data are processed and stored in a format known as trace files or chromatograms, which is then used to call the bases. The Phred software (Ewing et al. 1998a) produces more accurate base calls from these chromatograms than the software developed by ABI's own engineers and shipped with the machines (ABI PRISM Sequence Analysis Software, 1996). In addition to improved base-calling accuracy, the Phred program assigns an error probability to each base (Ewing and Green 1998). This **Phred score** has become the standard accuracy measurement for all types of DNA sequencing. Although the algorithms in the original Phred software are specific to the analysis of ABI chromatograms, the concept of a quality score that represents an accurate measure of error probability is applicable to all types of DNA sequencing.

Sanger sequencing and base calling by electrophoresis have several characteristic error patterns. The first few bases of each read are irregularly spaced and noisy because of anomalous migration of very short fragments and unreacted dye-primer or dye-terminator molecules as well as primer dimers and other artifacts. Toward the end of each read (beyond 500 bases), peaks become diffuse, lower in amplitude, and less evenly spaced as mass differences between successive fragments decrease. Another common problem is caused by the secondary structure of the fragments, commonly called compressions. This is most common in GC-rich sequences, which can form stable hairpins. Compressions cause uneven spacing of peaks in the chromatogram data, which often leads to insertion and deletion errors in base calling, for example, the sequence GG may be read as a single G peak or CC as a single C. Other sequence-specific error patterns have been noted specifically in the ABI chemistry, such as the decrease in signal amplitude for a G base following an A when using dye terminators (Parker et al. 1996).

Phred treats the chromatogram data as a set of equally spaced sine waves, representing a combination of all four of the bases, in a Fourier series (a frequency-modulated symmetric square wave). The amplitudes of all peaks for the four colors are normalized and scaled. Ideally, the fluorescent color signal with the maximum intensity at the predicted location for each wave in the series provides the correct base call. As peaks in the real data diverge from the ideal predicted locations and peak heights, a variety of empirically derived algorithms are used to correct the data, shifting observed peaks to match the nearest predicted peak location. "The full implementation is a complex, somewhat inelegant rule-based procedure, that has been arrived at empirically by progressively refining the algorithms on the basis of examining performance on particular data sets." (Ewing and Green 1998). Phred produced 40%–50% fewer base-calling errors (including both substitutions and insertions/deletions) per read than the ABI software.

The base-calling methods in the Phred software are highly specific to ABI chromatograms, but the methods for calculating the quality of sequence data have been applied to all DNA sequencing, including NGS technologies. The fundamental concept of Phred sequence quality scores is that an error probability is assigned to each base, which accurately reflects the chance of error in that base. In addition, the error probabilities have been designed to discriminate differences in accuracy within the high-quality regions of sequence reads. Phred error scores are reported as the log-transformed value of the error probability using the formula

$$q = -10 \times \log_{10}(p),$$

where p is the estimated error probability and q is the Phred quality score. Thus, a base with an error probability of 1/100 (99% accuracy) would receive a Phred score of 20, and an error probability of 1/1000 (99.9% accuracy) is equivalent to a Phred score of 30. This log transformation has the useful property that better quality bases have higher scores. The actual method for assigning Phred quality scores to bases in ABI sequence reads involves direct examination of the chromatogram files for properties such as peak spacing, ratio of uncalled to called bases, and peak resolution within windows of 3–7 bases. But more generally, the quality score must be predictive (computed without prior knowledge of the actual correct base at each position), and it must be valid—so that predicted scores correspond well with actual observed errors. Phred performs well by these criteria, with quality scores that are highly correlated with the frequency of observed errors in actual sequencing data.

Once Phred created base-by-base quality scores for sequence reads, it became possible to use quality as a criterion for sequence assembly software. Rather than trimming low-quality portions of sequence reads based on the frequency of ambiguous *N* base calls, trimming can be based on the average Phred score of the bases in a window. Typically, automatic trimming algorithms use a sliding window (i.e., 10–20) of bases and trim bases from both the 5′ and 3′ ends of each read until an average Phred score of adequate quality (i.e., ≥20) is reached. In an assembly of overlapping reads, quality scores can also be used to automatically choose the highest-quality bases to resolve conflicts between inconsistent base calls in different reads, rather than just relying on a strict "majority rule" system. Phred scores are also often used to provide an overall QC metric for an entire sequence or a set of sequences, such as average Phred score, or percentage of bases with Phred score above 30. Bases with Phred scores of 30 or above are generally considered to be of high quality, and values below Phred 20 are considered poor quality.

The creators of Phred also developed the Phrap program for assembly of fragments into contigs and the cross-match program to identify and remove vector sequence from the ends of sequence reads (Green 1994; de la Bastide and McCombie 2007). The Phrap program uses the Smith–Waterman algorithm (swat) to find matches between sequence reads and creates overlaps that can then be combined into contigs. Phrap also uses the concept of word hashing from heuristic alignment methods such as BLAST and FASTA to improve speed and efficiency. Given a minimum word size, Phrap only tests for alignment between reads that share a perfect match between words. Alignments are extended from each word match with matches contributing positive scores, mismatches and gaps incurring penalties. Parameters are set so that sequences with ~70% identity will have positive alignment scores. Reads are joined into contigs with the highest-scoring positive match. Phrap is designed to assemble a maximum of 64,000 reads with a maximum length of any

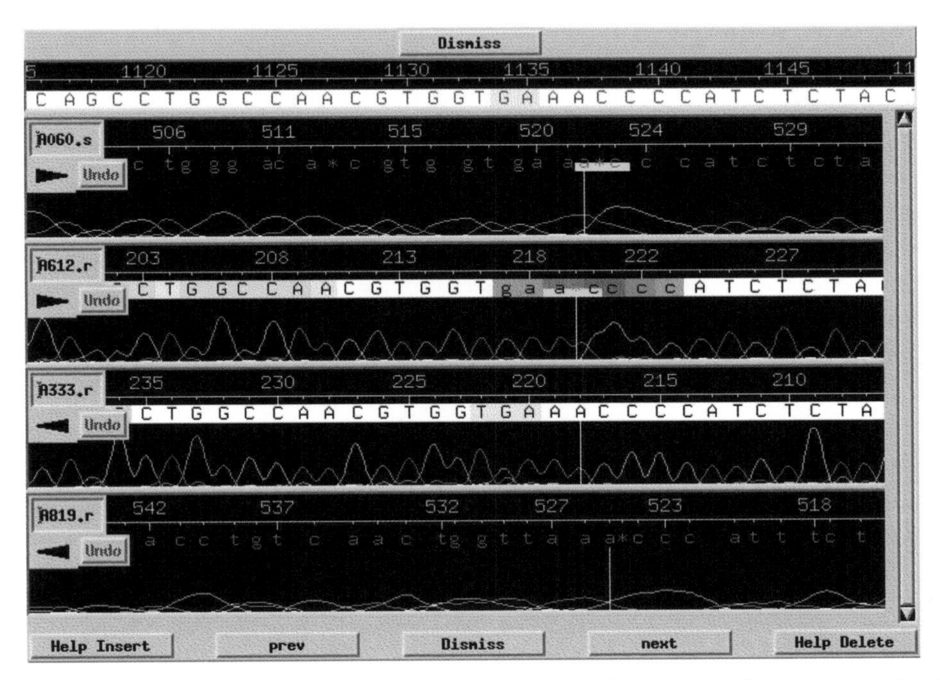

FIGURE 3. An X-Windows view of the Consed interactive contig editor showing chromatograms for three sequence reads.

single sequence of 64,000 bases, but it can be recompiled with the ".manyreads" and/or ".longreads" options to remove these limits. An important computational constraint of the Phrap algorithm is that each sequence read is compared with every other sequence read to find word matches, and then sequences with matching words are Smith–Waterman aligned. Therefore, the total amount of computing required by Phrap increases with the square of the number of reads, and it will be extremely computationally demanding to assemble millions of reads using this method.

Contigs assembled with Phrap can be viewed and edited with the Consed graphical editor (Gordon et al. 1998). This tool provides an interactive color display of overlapping reads, which allows the user to easily identify discrepancies in aligned base calls between different reads and possible misassemblies. To aid in resolving errors, base calls are shaded by Phred quality score, and it is also possible to directly view chromatograms aligned with the sequence reads and the consensus sequence. In regions where disagreements cannot be resolved, Consed provides tools to design new **sequencing primers** to obtain additional data. Consed can also be used to identify SNPs in a single genomic sequence as compared with a reference, or among sequences obtained from multiple individuals (Gordon 2003) (see Fig. 3).

REFERENCES

Altschul SF, Gish W, Miller W, Myers EW, Lipman DJ. 1990. Basic local alignment search tool. *J Mol Biol* **215:** 403–410.

Altschul SF, Madden TL, Schäffer AA, Zhang J, Zhang Z, Miller W, Lipman DJ. 1997. Gapped BLAST and PSI-BLAST: A new generation of protein database search programs. *Nucleic Acids Res* **25:** 3389–3402.

Bonfield JK, Staden R. 1995. The application of numerical estimates of base calling accuracy to DNA sequencing projects. *Nucleic Acids Res* **23:** 1406–1410.

Bonfield JK, Whitwham A. 2010. Gap5—Editing the billion fragment sequence assembly. *Bioinformatics* **26:** 1699–1703.

Bonfield JK, Smith K, Staden R. 1995. A new DNA sequence assembly program. *Nucleic Acids Res* **23:** 4992–4999.

Clayton J, Kedes L. 1982. GEL, a DNA sequencing project management system. *Nucleic Acids Res* **10:** 305–321.

Dayhoff MO, Schwartz R, Orcutt BC. 1978. A model of evolutionary change in proteins. In *Atlas of protein sequencing structure*, Vol. 5, Suppl. 3, pp. 345–358. National Biomedical Research Foundation, Silver Springs, MD.

de la Bastide M, McCombie WR. 2007. Assembling genomic DNA sequences with PHRAP. In *Current protocols in bioinformatics* (ed. Baxevanis AD, Davison DB), Chapter 11, unit 11.4. Wiley, New York.

Devereux J, Haeberli P, Smithies O. 1984. A comprehensive set of sequence analysis programs for the VAX. *Nucleic Acids Res* **12:** 387–395.

Eck RV, Dayhoff MO. 1966. *Atlas of protein sequence and structure.* National Biomedical Research Foundation, Silver Springs, MD.

Edman P, Högfeldt E, Sillén LG, Kinell P. 1950. Method for determination of the amino acid sequence in peptides. *Acta Chem Scand* **4:** 283–293.

Ewing B, Green P. 1998. Basecalling of automated sequencer traces using phred. II. Error probabilities. *Genome Res* **8:** 186–194.

Ewing B, Hillier L, Wendl M, Green P. 1998. Basecalling of automated sequencer traces using phred. I. Accuracy assessment. *Genome Res* **8:** 175–185.

Fitch WM. 1966. Evidence suggesting a partial, internal duplication in the ancestral gene for heme-containing globins. *J Mol Biol* **16:** 17–27.

Gordon D. 2003. Viewing and editing assembled sequences using Consed. *Curr Protoc Bioinformatics* **2:** 11.2.1–11.2.43.

Gordon D, Abajian C, Green P. 1998. Consed: A graphical tool for sequence finishing. *Genome Res* **8:** 195–202.

Gumbel EJ. 1958. *Statistics of extremes.* Columbia University Press, New York.

Karlin S, Altschul SF. 1990. Methods for assessing the statistical significance of molecular sequence features by using general scoring schemes. *Proc Natl Acad Sci* **87:** 2264–2268.

Lawrence CB, Solovyev VV. 1994. Assignment of position-specific error probability to primary DNA sequence data. *Nucleic Acids Res* **22:** 1272–1280.

Lipman DJ, Pearson WR. 1985. Rapid and sensitive protein similarity searches. *Science* **227:** 1435–1441.

Lipshutz RJ, Taverner F, Hennessy K, Hartzell G, Davis R. 1994. DNA sequence confidence estimation. *Genomics* **19:** 417–424.

Miller MJ, Powell JI. 1994. A quantitative comparison of DNA sequence assembly programs. *J Comput Biol* **1:** 257–269.

Needleman SB, Wunsch CD. 1970. A general method applicable to the search for similarities in the amino acid sequence of two proteins. *J Molec Biol* **48:** 443–453.

NIH. 1990. Understanding our genetic inheritance: The United States Human Genome Project: The first five years: Fiscal years 1991–1995. DOE/ER-0452P, NIH Publication No. 90-1590; http://www.genome.gov/10001477.

Nirenberg M, Leder P, Bernfield M, Brimacombe R, Trupin J, Rottman F, O'Neal C. 1965. RNA codewords and protein synthesis. VII. On the general nature of the RNA code. *Proc Natl Acad Sci* **53:** 1161–1168.

Parker LT, Zakeri H, Deng Q, Spurgeon S, Kwok PY, Nickerson DA. 1996. AmpliTaq DNA polymerase, FS dye-terminator sequencing: Analysis of peak height patterns. *BioTechniques* **21:** 694–699.

Pearson WR, Lipman DJ. 1988. Improved tools for biological sequence comparison. *Proc Natl Acad Sci* **85:** 2444–2448.

Sanger F, Tuppy H. 1951a. The amino-acid sequence in the phenylalanyl chain of insulin. 1. The identification of lower peptides from partial hydrolysates. *Biochem J* **49:** 463–481.

Sanger F, Tuppy H. 1951b. The amino-acid sequence in the phenylalanyl chain of insulin. 2. The investigation of peptides from enzymic hydrolysates. *Biochem J* **49:** 481–490.

Sankoff D. 1972. Matching sequences under deletion/insertion constraints. *Proc Natl Acad Sci* **69:** 4–6.

Smith TF, Waterman MS. 1981. Identification of common molecular subsequences. *J Mol Biol* **147:** 195–197.

Staden R. 1977. Sequence data handling by computer. *Nucleic Acids Res* **4:** 4037–4051.

Staden R. 1979. A strategy of DNA sequencing employing computer programs. *Nucleic Acids Res* **6:** 2601–2610.

Staden R, Beal K, Bonfield JK. 1998. The Staden Package. Computer methods in molecular biology. In *bioinformatics methods and protocols* (ed. Misener S, Krawetz SA), Vol. 132, pp. 115–130. Humana Press, Totowa, NJ.

WWW RESOURCES

http://bozeman.mbt.washington.edu/phredphrap/phrap.html Green P. 1994. Documentation for PHRAP. Genome Center, University of Washington, Seattle. (Accessed on 5/15/2011.)

http://www.ncbi.nlm.nih.gov/BLAST/tutorial/Altschul-1.html Altschul S. The statistics of sequence similarity scores. NCBI BLAST tutorial. National Center for Biotechnology Information, National Library of Medicine. (Accessed on 5/15/2011.)

3

Visualization of Next-Generation Sequencing Data

Phillip Ross Smith, Kranti Konganti, and Stuart M. Brown

Large, complex data sets such as NGS are very difficult to understand even when relatively straightforward analyses provide ready access to the information they contain. Techniques of data visualization often provide a powerful way to pick out important features in the data, such as correlations and skew, that are not readily apparent and allow an investigator to devise new data analytic approaches that permit their quantification.

This chapter explores data visualization in the context of **ChIP-seq**, **RNA-seq**, and other deep sequencing experiments. The data that are obtained are voluminous, from millions to billions of data records that are delivered to the investigator in huge text files or spreadsheets. Direct numerical analysis of such data requires **high-performance computing** facilities and/or accelerated **algorithms**. However, visualization in the context of appropriate analytics can be very revealing of important biology because the experimental information can be viewed in an appropriate context.

We begin by exploring the data formats used for data representation in NGS. We then look at the visualization challenges posed by each of the main types of NGS experimental applications including ChIP-seq, RNA-seq, variant discovery, deep sequencing of **amplicons**, and **de novo assembly**. To conclude, we will try to outline the thinking that an investigator might want to follow when considering a decision to choose among existing visualization software or to create a new visualization tool required to fill local needs.

HISTORY

One of the first software tools developed to assist investigators analyze sequencing data was the Staden package, the first elements of which were authored in 1977 and which was finally published in 1998 (Staden et al. 1998). Its particular value

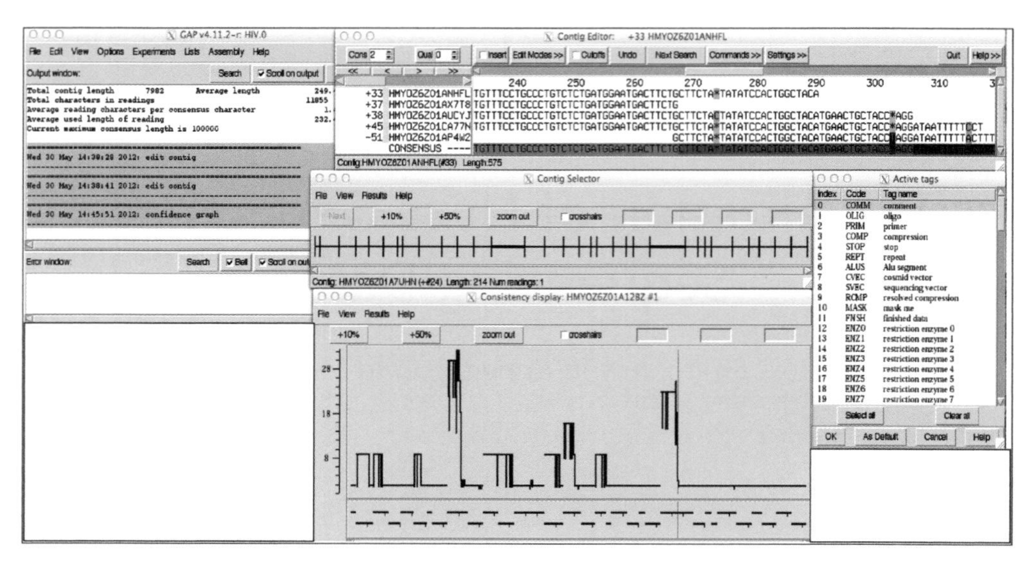

FIGURE 1. The Staden GAP4 package provides a visual interface for assembly and editing of DNA sequence **contigs**.

is that it was first to indicate the central role of the computer in managing the tasks of nucleic acid **sequence assembly** and visualization. (See Fig. 1.) The package has achieved very high usage and is still actively maintained as a project on the SourceForge site (http://staden.sourceforge.net/).

Data Representation

The raw data output from sequencing machines contains all of the information that the machine can provide. These data formats are proprietary and depend on the manufacturer of the sequencing equipment. Among these is the **SFF format** (Standard Flowgram Format), which is the output of choice of the **Roche 454 sequencing** machines. There are also qseq.txt and .bcl files created by Illumina, and many variants of **FASTQ formats** (sequence and quality) adopted by different vendors and modified across various versions of their software.

These output formats are very voluminous because they contain the sequence data, quality measures for the entire read and for individual bases that are called, identifiers for read, plate location, direction, pairing, and other data that may be needed to resolve **sequence alignment** issues as the data are assembled.

Filtering

Raw data are generally very voluminous, and most analyses only require a subset of the fields provided in each record, or a quality filtered subset of all reads collected by the machine. Filtering methods are used to remove the unneeded fields and to produce a more compact representation of the data.

Genomic Position

In many deep-sequencing projects, short **sequence reads** (tags) are aligned to a **reference genome**. In this case, the actual sequence can be fully defined by the assembly ID of the reference genome (i.e., hg18), the chromosome, the location of the start position on that chromosome, the length of the sequence string, and any differences between the reference genome and the read. These differences are recorded as mismatches by the **alignment algorithm** (see Chapter 4), which may subsequently be identified as **sequence variants** by a **variant detection** algorithm (see Chapter 8).

Data Formats

FASTA and FASTQ

Historically, the **FASTA** (http://en.wikipedia.org/wiki/FASTA_format) and **FASTQ** (http://en.wikipedia.org/wiki/FASTQ_format) **sequence data formats** have played pivotal roles in analysis. These are text files with a single header line beginning with an identifying start character, ">" in the case of FASTA and "@" in the case of FASTQ.

FASTA files store simple sequence data (either nucleic acid or protein), usually fewer than 80 characters in a line. In the case of a nucleic acid sequence, there are 13 additional characters aside from the usual A, C, G, T, U that code for uncertainties in the sequence identifications (see http://en.wikipedia.org/wiki/FASTA_format). FASTQ files store sequence data along with a corresponding quality score (http://en.wikipedia.org/wiki/FASTQ_format). The FASTQ format was not widely used for **Sanger sequence** data but has become the standard data storage and exchange format for NGS data. Data stored in FASTA and FASTQ formats can be processed by a broad range of software packages. They are the most common data exchange formats for bioinformatics software and the default download format for all public sequence databases.

There are many additional data formats that are used for NGS data. Depending on the use to which the data are put, these vary significantly in complexity and in the data elements stored in the file. If you are a beginner, interested in getting a quick look at the data you have collected, familiarity with the simplest formats will probably suffice in the short run: these are ALN and BED and are described below. Table 1 shows a compilation of most of the common file formats at the time of final revision of this chapter. The table contains links to the sources of the definitive descriptions of these formats. You should be aware that the definitions of the file formats can change and,

TABLE 1. Next-generation sequencing data file types

Name	Web description	Main package	Type
ALN	http://www.biostat.jhsph.edu/~hji/cisgenome/index_files/tutorial_seqpeak.htm	CisGenome	TSV
BED	http://genome.ucsc.edu/FAQ/FAQformat.html#format1	UCSC	TSV
FASTA	http://en.wikipedia.org/wiki/FASTA_format	FASTA	TSV
FASTQ	http://en.wikipedia.org/wiki/FASTQ_format	SAMTools	TSV
GFF3	http://gmod.org/wiki/GFF3	Gmod	TSV
SAM/BAM	http://samtools.sourceforge.net/SAM-1.3.pdf	SAMTools	TSV/Binary
USEQ	http://useq.sourceforge.net/useqArchiveFormat.html	IGB	Binary
WIG	http://genome.ucsc.edu/goldenPath/help/wiggle.html	UCSC	TSV

with a few exceptions, are not necessarily subject to any formal review and standard setting. In addition, software authors may modify these formats in inconsistent ways to contain data they need for their own program packages. Let the user beware.

ALN

ALN is a simple, straightforward genomic position format that is used as the principal data format for the "CisGenome" package, among others. It is a flat file, one record for each tag location with three tab-delimited fields as follows:

```
chromosome[tab]location[tab]strand
```

An example might be:

```
chr10 1021346 +
chr10 1021456 –
etc. ...
```

In an ALN file, the location is measured on the "+" strand irrespective of the read direction ("+" or "–"). Because there is a single field for location, the ALN format assumes that all sequence reads are the same length, and this length must be defined external to the file.

The nomenclature for the chromosomes should be consistent among the various data sources that you want to use. A strand can be denoted by +/– or F/R.

Some software authors "load" the records with additional values. We discuss this below in the chapter.

BED

A record in the BED format is similar to one in ALN. In its simplest form, it looks like this:

```
chromosome[tab]StartLoc[tab]EndLoc ...
```

There are, however, nine additional fields that can be optionally included in a BED-formatted record. These are needed because **BED files** are used primarily to contain information needed for data display in the UCSC Genome Browser. Beginning with field 4, the names of these fields in a BED record are: 4-name, 5-score, 6-strand, 7-thickStart, 8-thickEnd, 9-itemRgb, 10-blockCount, 11-blockSizes, 12-blockStarts. The order of these optional fields is required: the lower-numbered fields must always be populated if a higher-numbered field is used. A record in BED format that contains the minimal information in an ALN file must have at least six fields if the "strand" information is required for display or analysis. More information as to the detailed meaning and use of these fields can be obtained from the reference in Table 1. Once again, it is important to note that "BED" files defined in the context of software other than the UCSC Genome Browser may not be identical or compatible (http://qed.princeton.edu/main/Wiggle_(BED)_files).

Wiggle (WIG)

The WIG format is used in the UCSC Genome Browser website as a method to store information for display in a compact way. The WIG format defines display parameters for intervals of the genome, not for individual sequence reads. It is often used to display integrated information or analysis results. It works best for the "display of dense, continuous data such as GC percent, probability scores, and transcriptome data." Individuals who are heavy users of the UCSC Genome Browser will find this format of value for the storage and exchange of data for display, but it isn't particularly useful for storing data as part of an analysis pipeline. The UCSC web page (http://genome.ucsc.edu/goldenPath/help/wiggle.html) provides more regarding this format.

GFF3

GFF3 is the Generic Feature Format, version 3 (http://gmod.org/wiki/GFF3), which is a newer version than the originally proposed GFF format. GFF3 is a format native to the GMOD package in which genomic sequence features are stored in a tab-delimited flat file. Complete gene structures with annotations are represented in a nested pattern, which stores information regarding the location of these features on the chromosome. A genomic sequence and its annotation in GFF3 format is mainly used in displaying its features in graphical format in Generic Genome Browser (GBrowse). Readers are encouraged to learn about the GFF3 format from The Sequence Ontology Project (http://www.sequenceontology.org/gff3.shtml).

SAM and BAM

SAM stands for "Sequence Alignment/Map," the format defined formally by the "SAM Format Specification Working Group," which completed v1.3-r882 of the specification in December 2010 (the definition is in continual revision). The SAM file format is complex and its details are best read about on the website given in Table 1. SAM is used to store sequence reads aligned to a reference genome.

BAM files are binary files that consist of indexed, block-compressed, SAM data constructed so that the index can provide fast, direct access to any part of the file. This format is therefore ideal for many display tasks, particularly using the UCSC Genome Browser, because only the data needed for the display of the currently selected genomic interval need be uploaded, resulting in a significant increase in display speed. The format of BAM files is strictly defined and is machine independent (http://samtools.sourceforge.net/SAM1.pdf).

Useq

Finally, we mention "Useq," which is a binary format allowing a directory of genomic data to be zip-compressed. Using this scheme, very large amounts of preindexed genomic information can be archived and transmitted. The web page describing it indicates the packages that can access these data directly and provides details as to the organization of the data within the archive (http://useq.sourceforge.net/useq ArchiveFormat.html).

SAMtools

Different data formats are needed for optimal use of different tools. Fortunately, there are several easy-to-implement toolboxes, such as SAMtools (http://samtools.sourceforge.net/), with software to interconvert between data formats, so that data can be made available for packages that do not read them natively.

Choice of Visualization Software

Visualization tools need to be convenient to use, which means, practically, that they need to run on the computer typically used for work at the desktop. Right now, investigators will almost certainly have a PC running a flavor of MS Windows, a

Macintosh, or a Linux workstation. Because most laboratories are likely to have investigators using all of these platforms, there is a benefit to have software serving a common task that is capable of running on all three platforms, so that everyone can participate equally in data analysis and the inspection of the results.

The software that investigators need may run either locally, on the desktop, or remotely on some other platform. There are many options for the latter, the obvious example being software that is accessed via the Web, such as the UCSC Genome Browser.

There are three ways, in general, to create cross-platform software that is to run locally. The obvious method is to create functionally identical packages built for each of the platforms individually. This usually produces the fastest versions of the software, but it can significantly slow development if each platform has different requirements and will add dramatically to the cost and time for maintenance as the developer is forced to keep up with system software upgrades for each platform.

The second method is to use a scripting language or a cross-platform software toolkit (such as Perl or the R language) to build the software. In this case, the software code is identical across platforms, which simplifies development and deployment. However, the scripted software requires platform-specific libraries in order to execute. Because these libraries are, usually, highly standardized and optimized, the software author does not need to worry about low-level implementation issues. The price paid is that the library tools are generic and may not be tailored to the specific needs of the package, thus overall performance may be mediocre. Performance is a critical issue for NGS analysis packages because the data are usually very large.

The final method is to write the software in Java. The big advantage of Java is that it is a true cross-platform system. As with the scripting method, a high-performance Java Virtual Machine needs to be installed on the platform where the software is to run, but Java is now very well established, and it is hard to find a desktop where Java is not available. The drawback of software built in Java is that the Virtual Machine often requires more available RAM than executing similar operations in code that is optimized to run natively on the desktop computer's own operating system.

The bottom line is that software built native to the platform to be used has the potential to run faster and to be able to take advantage of particular features of the specific machine being used, such as high-performance storage, multiple CPUs, very large memory, and software optimizations at the CPU level. These potentials are no guarantee that they will be exploited, and thus in many cases, scripted software and software implemented in Java will have very competitive performance. Some investigators who write their own software do so initially in a scripting language, such as Perl, then port it to native code for large-scale production only if the Perl version is unacceptably slow. In this fashion, the relative simplicity of the scripting language speeds algorithm development, and natively built software is created only when and if it is needed.

The Analytic Flow

The preparation of data for analysis follows three steps: sequencing machine generation of the raw sequence reads; the alignment of these reads to a reference or to each other; and the visualization of the aligned sequence reads and their interpretation in the light of the relevant biology.

Raw Sequence

There are numerous sequencing machines designed to perform massive sequencing tasks, and the list is changing constantly. In Chapter 1 we list a number of these machines, and a more complete review is available (Rizzo and Buck 2012). The details of the construction, chemistry, and data recording methods are specific to each machine, but essentially, each of these machines works by conducting cycles of chain elongation that copy many individual template DNA molecules. Imaging (or ion detection) reveals the fragments that were elongated with each specific base, and the cycle is repeated. Analysis of these images allows the sequence to be recorded. DNA fragments that fail to produce high-quality base calls are flagged and may be removed from the raw sequence data file. Data files produced by the machine are voluminous, but the data records all contain three essential pieces of information: an ID defining the specimen being analyzed, the sequence of the fragment, and a quality estimate for each base call.

Alignment

The next step in a project is the **alignment** of the fragments. Alignment can be to a **reference sequence** or fragment to fragment (de novo assembly). Illumina and ABI/SOLiD machines produce **short reads** (25–100 bases), whereas the Roche 454 sequencer produces reads that are typically 200–800 bases long.

The alignment is usually performed by software provided by the manufacturers of the sequencing machine, with standard, built-in alignment parameters. However, fragment alignment can be undertaken by the investigator him/herself if desired. Chapter 4 describes some of the alignment packages that are available for use. Alignment (or de novo assembly) is usually the most computationally demanding aspect of the NGS analytic workflow. Laboratories that own NGS machines typically use dedicated high-performance computational clusters with large amounts of RAM and access to extremely high-capacity, high-speed data storage systems (see Chapter 12).

The results of the alignment process are data that consist (minimally) of a chromosome (or scaffold), the alignment location on that chromosome, the strand, and an alignment quality estimate. By their nature, these data are very voluminous, posing significant issues for storage, archiving, and for manipulation in subsequent analytical steps. For this reason, "raw" alignment data records will usually need to be filtered to retain only those data fields that are needed for analysis.

ChIP-seq

The simplest file format retaining the minimal information needed for the analysis of ChIP-seq data is the ALN format, which retains only chromosome, location, and strand. Base-call quality, alignment quality, and sequence mismatch information are generally not used for ChIP-seq analysis (see Chapter 9). The SAM format is much richer supporting many different analyses, but with more fields in each record required for this to be possible. However, the BAM format allows these data to be both compressed and indexed, so that there is efficient random access to subsections of the data as required.

RNA-seq

RNA-seq data are very similar to ChIP-seq data, but instead of sequencing DNA fragments, mRNA fragments are sequenced. This allows an investigator to study the transcriptome of cells of interest, which provides direct information regarding actual protein synthesis at the time that the samples were prepared for analysis (see Chapter 10).

As in ChIP-seq, tag sequences are aligned with a reference genome. But because the tags are RNA, they align, to a large extent, only with the **exons** in the genome, not the **introns** and not in regions of the genome that do not code for genes. In practice, of course, fragments are found that bind in these regions, and these data can be very valuable.

They are particularly valuable if an investigator is studying tissues from a tumor. In this case, the transcribed mRNA can come from genes that have been formed by transpositions of genomic DNA between chromosomes. RNA-seq allows these transpositions to be identified because one end of a RNA tag will map to a location on one chromosome, and the other end will map to a remote location, often on a different chromosome. Interpretation of these data is difficult and time-consuming, and access to an appropriate visualization tool can be very helpful.

Visualization

The optimum choice of visualization tool will depend on the biological problem being studied. We begin by looking at a simple task, that of identifying binding sites of transcription factors to DNA using ChIP-seq.

The purpose of a ChIP-seq experiment is to find parts of the genome where proteins bind preferentially. In a simple example, a transcription factor is bound to the

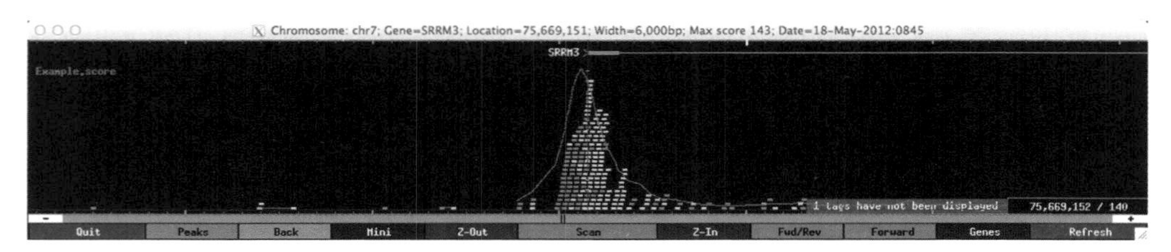

FIGURE 2. A tag peak located using a score computed similarly to the MACS algorithm and visualized using a simple display tool developed in our laboratory here at NYULMC.

genome and is cross-linked to stabilize the DNA–protein interaction. The DNA is fragmented, and the DNA–protein complex is immunoprecipitated using an antibody specific to the transcription factor. The DNA fragments are released from cross-links to the bound protein and sequenced. The DNA sequences are then aligned to the reference genome. The aligned sequence reads are called "tags."

Intuitively, it is clear that the tags obtained should cluster in the regions of the genome where the transcription factor binds. When the experiment is visualized, the investigator needs to see "peaks" consisting of clusters of tags on the DNA strands that indicate preferred binding.

The "problem" is not so much the visualization of the peaks, as defining how and where to look for them. It is very clear that even if there are thousands of peaks, it is unlikely that looking in on a single stretch of DNA/RNA will reveal much.

Whereas deciding whether a given peak is significant is a task requiring sophisticated analysis software (see Chapter 9), it is easy and very quick to generate a "score" from the binding locations that can be used to pinpoint the locations on the DNA strands where tags are clustered. The investigator can then step through the locations of local maxima in the score, confident that the large majority of peaks will have been inspected (http://www.ncbi.nlm.nih.gov/projects/geo/query/acc.cgi?acc=GSE13047).

In Figure 2 we show a tag peak located using a score computed similarly to the MACS algorithm and visualized using a simple display tool developed in our laboratory here at the New York University Langone Medical Center (NYULMC). Other packages can provide similar results. The tool used most commonly in the community is probably the UCSC Genome Browser.

How to Choose Visualization Tools

To decide how best to visualize data, you need to know some basic facts. First, where are the data stored? Considerable processing is needed to prepare data for analysis and display, which, for the most part will be performed remotely on the analysis cluster associated with the sequencing laboratory. The decision to be made is whether to move the data files to the local workstation or to access them over a network connection.

What are the capabilities of the local workstation? Key factors are local disk capacity, RAM, CPU power (including the number of processor cores), the workstation operating system, and the design of the visualization software. Is the workstation able to display data computed remotely?

In general, nearly all data analysis occurs as part of a pipeline, thus the "best" data analysis choices are those that optimize the efficiency of the process overall, minimizing the minor "waits" for data to transfer or display and facilitating the move to the next steps in the analysis. From the standpoint of data visualization, "best" choices allow an image to appear on the screen most quickly. Usability and the learning curve are also important factors. Software that is slow and cumbersome to use consumes time constantly throughout all analyses. Software that requires substantial learning to operate requires an initial investment of time. Software that is sophisticated and complex to operate (or that provides a very open-ended toolkit) may create a premium on expert users, so that the analysis of a single data set will produce different results in the hands of different data analysts.

Another key issue is the costs associated with software licenses, the time taken to install packages on the target machines (a local workstation or a more central server), and the costs of software maintenance. Local software needs to be maintained, and the maintainer is usually the owner. Time spent installing and updating software is time not spent working with the data and moving a project forward. Senior investigators need to be aware of these types of trade-offs and encourage efficient software choices for members of a research team.

Software Comparisons

A reader of this chapter will probably be on the point of making choices for the best software to meet a project need. It is almost inevitable that the options available at the time you read this will be different from those that seemed most attractive at the time we prepared this material. Your goal should be to understand the strengths of each package, to decide whether these are key strengths for the project, and then to look for them in the latest software offerings. These include a subset of the file formats that can be read natively, whether the software needs to be run remotely or locally and therefore installed on the workstation, the display mode, the amount of data that can be displayed on the screen at one time, and the time to display a peak on the screen from a suitable input file.

UCSC Genome Browser

The UCSC Genome Browser is a web-based genome browser that is very widely used for visualization of all types of location-specific annotation on human and model organism genomes. UCSC was commonly used to visualize early versions of NGS

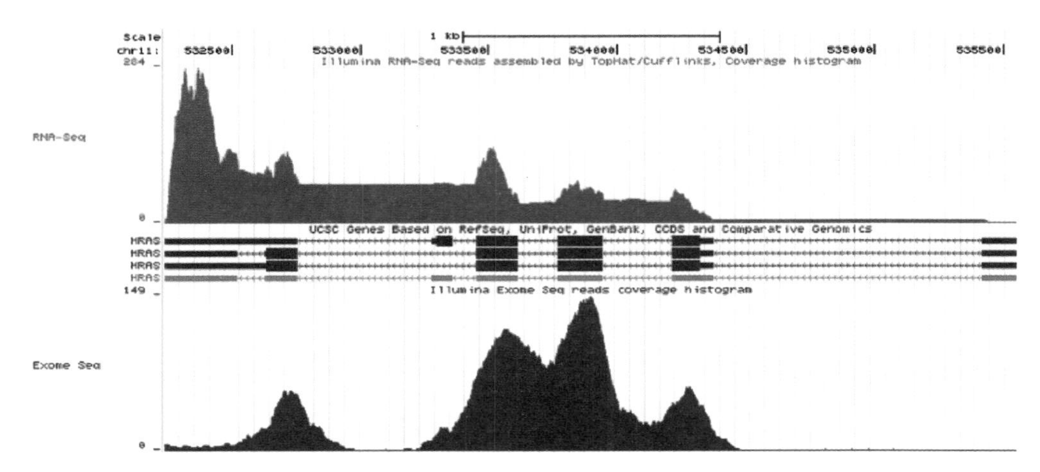

FIGURE 3. UCSC Genome Browser display of coverage graphs of Illumina RNA-seq reads assembled into transcripts by the TopHat/Cufflinks package (green) and Illumina Exome sequencing reads (blue) at the *HRAS* gene.

data (after alignment to a reference genome) as "custom tracks" uploaded by the user. Because in Illumina and SOLiD the sequence data are on the order of tens of gigabase pairs, the UCSC Genome Browser has implemented compressed binary file formats known as BigBed and BigWig to allow large file transfer by the remote URL method.

The key advantages of the UCSC browser include the huge number of different annotation tracks optionally available to the user for display (**GenBank** sequences, predicted gene models, gene expression experiments, gene function, sequence variants, comparative genomics, chromatin structure, regulatory motifs, etc.) and the instant availability of the resource on the Web. The user must convert their data from BAM format into BigWig in order to view a **coverage** graph, as shown in Figure 3.

Illumina GenomeStudio

GenomeStudio (http://www.illumina.com/software/genomestudio_software.ilmn) is commercial software from Illumina that is sold to facilitate the analysis of data collected on their sequencing platforms. The package is built around modules that can allow the investigator to tackle all of the major challenges in nucleic acid sequencing research (the modules include the DNA Sequencing Module, the RNA Sequencing Module, the ChIP Sequencing Module, the Genotyping Module, the Profile miRNA expression, the Methylation Module, and the Protein Analysis Module). GenomeStudio runs on the Windows platform. Owing to the very large data sets that it integrates,

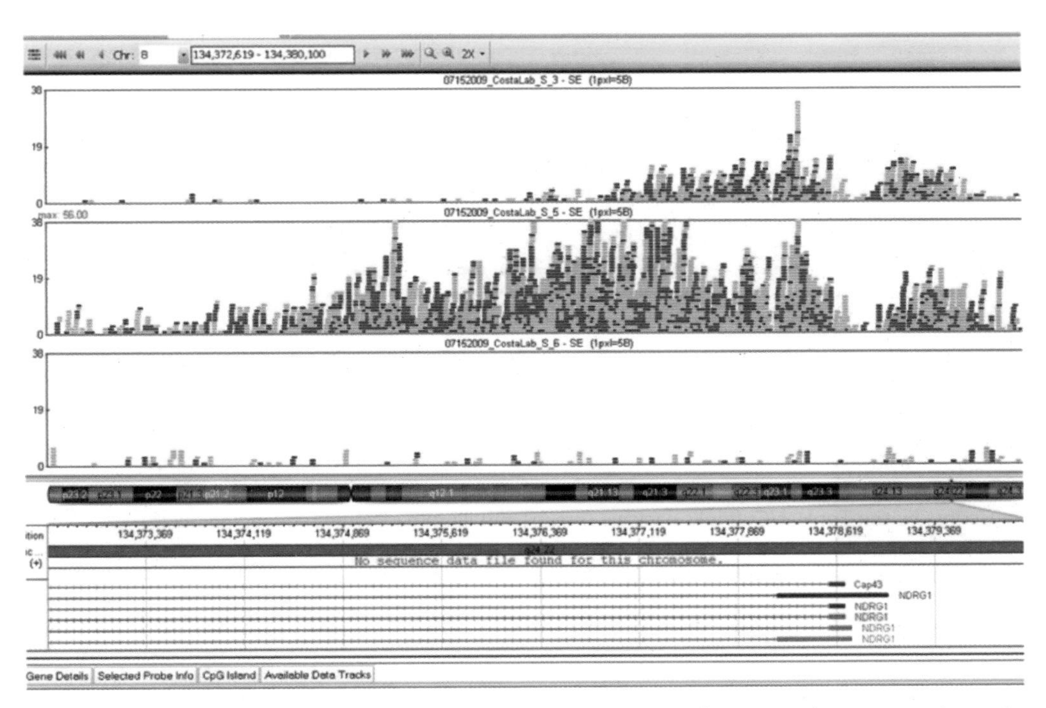

FIGURE 4. A screenshot of Illumina Genome Studio visualizing reads from two ChIP-seq samples and a genomic DNA control, aligned to the reference human genome near the 5′ end of gene *NRDG1*.

at least 8 GB of RAM and a multicore CPU are required for good performance. GenomeStudio can read and export data from all of the major data formats and can therefore be integrated into the workflow that an investigator chooses (see Fig. 4).

Mauve

Mauve (http://asap.ahabs.wisc.edu/software/mauve/) is a package designed to help construct multiple genome alignments (Darling et al. 2010). It permits the project to be followed from its earliest stages through to the final characterization of the alignment. The package is written in C and can be downloaded and compiled from source if desired for Windows, UNIX, and Mac OSX platforms (see Chapter 6 for a more detailed discussion of Mauve).

newbler, Amplicon Variant Analyzer

newbler (http://contig.wordpress.com/2010/02/09/how-newbler-works/) (formally known as GS de novo Assembler) is a package developed for de novo assembly from data obtained from the Roche 454 Sequencer. Amplicon Variant Analyzer is a software package developed by 454 for **multiple alignment** and visualization of

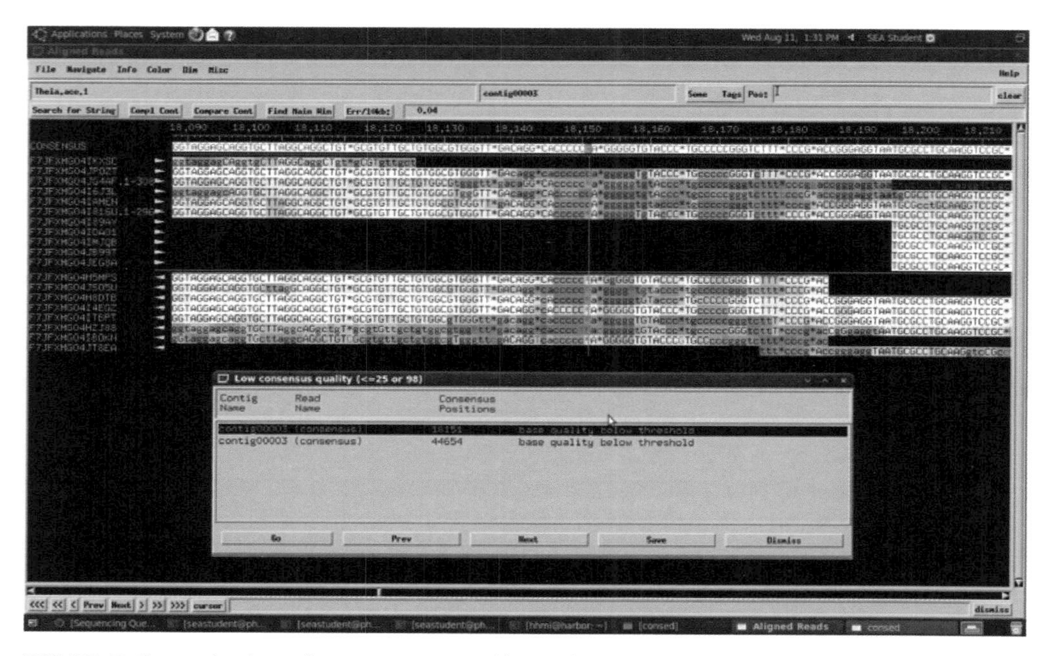

FIGURE 5. Base-pair view of a genome assembly in the 454 newbler application. Note the strong resemblance to the Staden Gap package.

very deep sequencing of defined **sequence fragments** (PCR amplicons). It has been used to identify rare mutations in in vitro populations of HIV. Roche 454 software runs on Linux and other UNIX platforms. It is freely available to Roche customers (see Fig. 5).

Integrative Genomics Viewer (IGV)

The Integrative Genomics Viewer (IGV) is free Java-based stand-alone desktop software developed at the Broad Institute (Robinson et al. 2011). It can visualize NGS data in a variety of native file formats (FASTA, FASTQ, SAM, and BAM) as well as **microarray** gene expression and genotyping data. As Java software, it is easily installed by a quick download on any computing platform, and it accesses data stored on a local workstation or remotely mounted network drive. Reference genomes and genome annotations must be manually installed. Visualization of large data sets requires substantial amounts of RAM on the local workstation (see Fig. 6).

GenomeView

An important new addition to the visualization packages for next-generation standalone genome browser is GenomeView. The package provides numerous important features including the interactive visualization of sequences, annotation, multiple

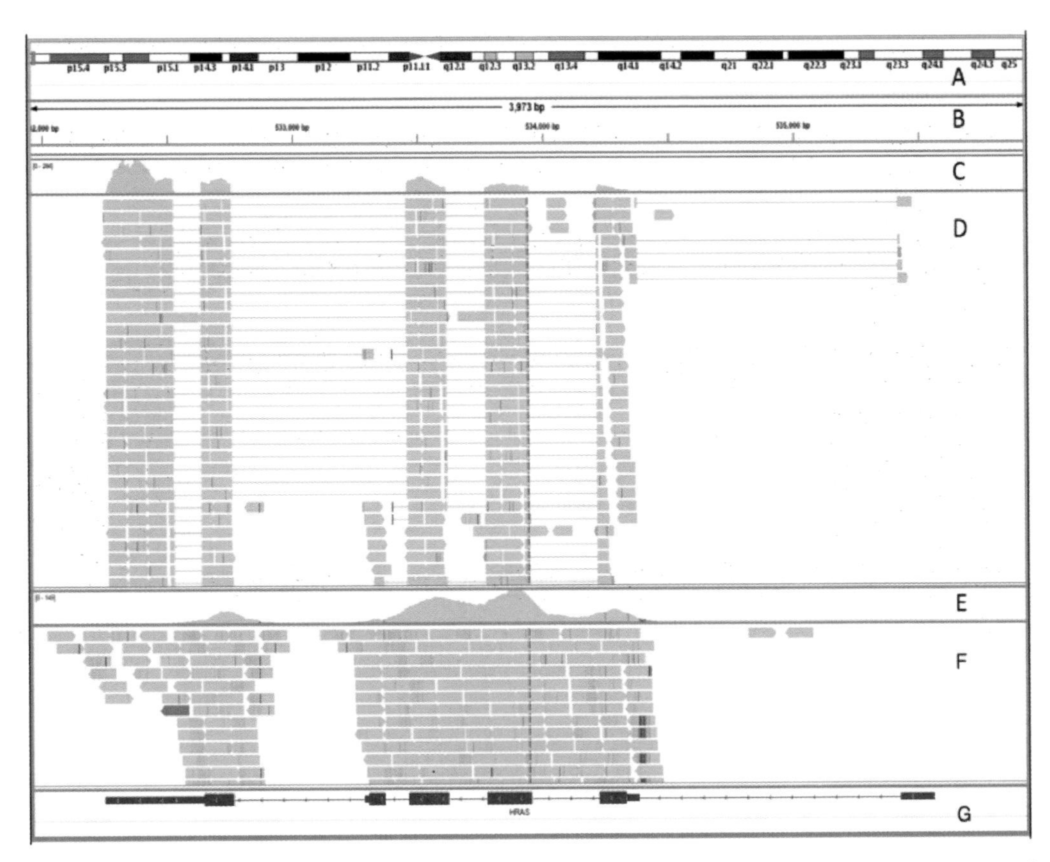

FIGURE 6. This is the main window of the IGV from the Broad Institute. (*A*) Chromosome overview of human genome hg19, Chr11 is shown here. (*B*) The chromosome region currently visualized. (*C*) IGV showing the coverage of TopHat/Cufflinks assembled transcripts in **SAM/BAM** format. (*D*) The RNA-seq reads aligned to the reference genome with thin lines indicating reads spill across introns. (*E*) Coverage map of a second set of RNA-seq reads mapped on the reference genome. (*F*) RNA-seq reads aligned to the reference genome. (*G*) Compact view of *HRAS* RefSeq gene model.

alignments, syntenic mappings, and short read alignments. Support has been provided for many standard file formats, and a plug-in system allows new functionality to be added as user needs evolve.

The popularity of this package has increased significantly in the last six months, as its enhanced visualization features have become well known. Important among these are the fact that BAM files can be read and written, and the user's ability to zoom rapidly and smoothly from the level of a chromosome down to individual bases in a multiple alignment analysis. The improvement in usability plus the fact that a large-memory version of the package is available largely remove any limitations to the projects that can be tackled using state-of-the-art methods.

Development of the package was initiated by Thomas Abeel in 2008 at the BSB group at VIB and now continues at the Broad Institute, his current institution. The

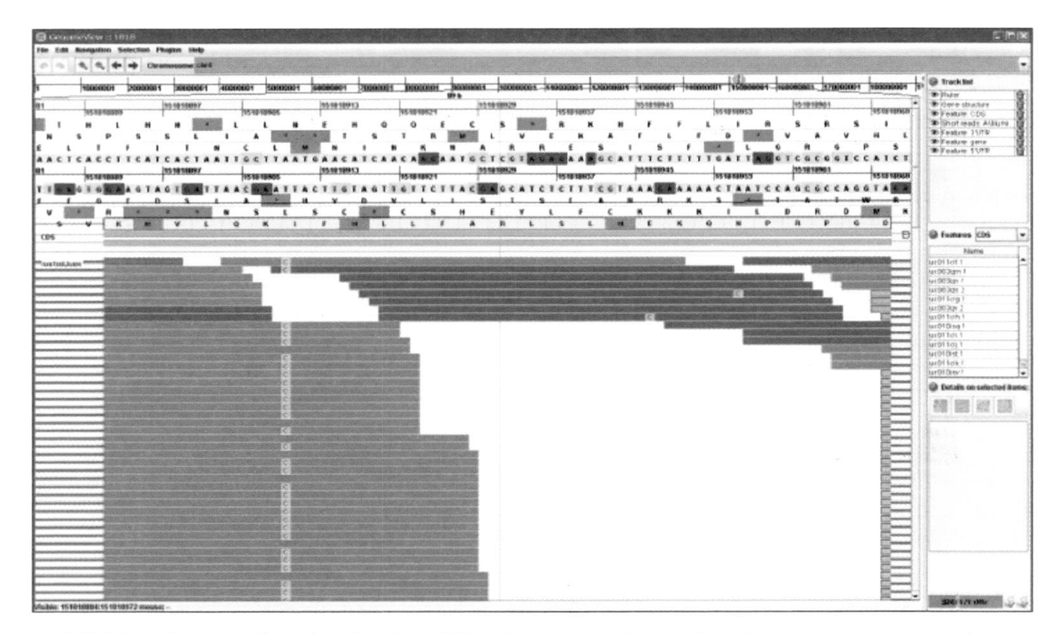

FIGURE 7. GenomeView visualization of **Illumina sequencing** reads to inspect a sequence variant.

package is written in Java and can be initiated from a local copy of the software or using a "webstart" from http://www.genomeview.org, the homepage of the package (see Fig. 7).

Use of GBrowse at a Sequencing Core facility

The Generic Genome Browser (GBrowse) is one of the software components from the GMOD (Generic Model Organism Database) community. It was developed by Lincoln Stein and coworkers (Stein et al. 2002) as a web-based genome visualization tool and therefore is platform-independent software. It is widely used by most of the model organism genome sequencing and annotation projects such as WormBase (http://wormbase.org), FlyBase (http://flybase.org), The *Arabidopsis* Information Resource (http://www.arabidopsis.org), and the like. GBrowse is a scalable web application to efficiently view large amounts of sequence information in a user-friendly graphical format. Standard reference genomes (human genome hg19, mouse genome mm9, etc.) can be stored on the server and used to annotate any uploaded data file. Gene structures (introns, exons, UTRs, etc.), microarray expression data, RNA-seq data, etc. are represented in uniquely identifiable images called *glyphs* and users can search for the landmark of their interest and browse by scrolling and zooming around the sequence tracks. Additional tracks that provide summary annotation for experimental samples can be computed automatically by standard GBrowse tools (GC content, 6-frame translation), or uploaded as additional GFF files.

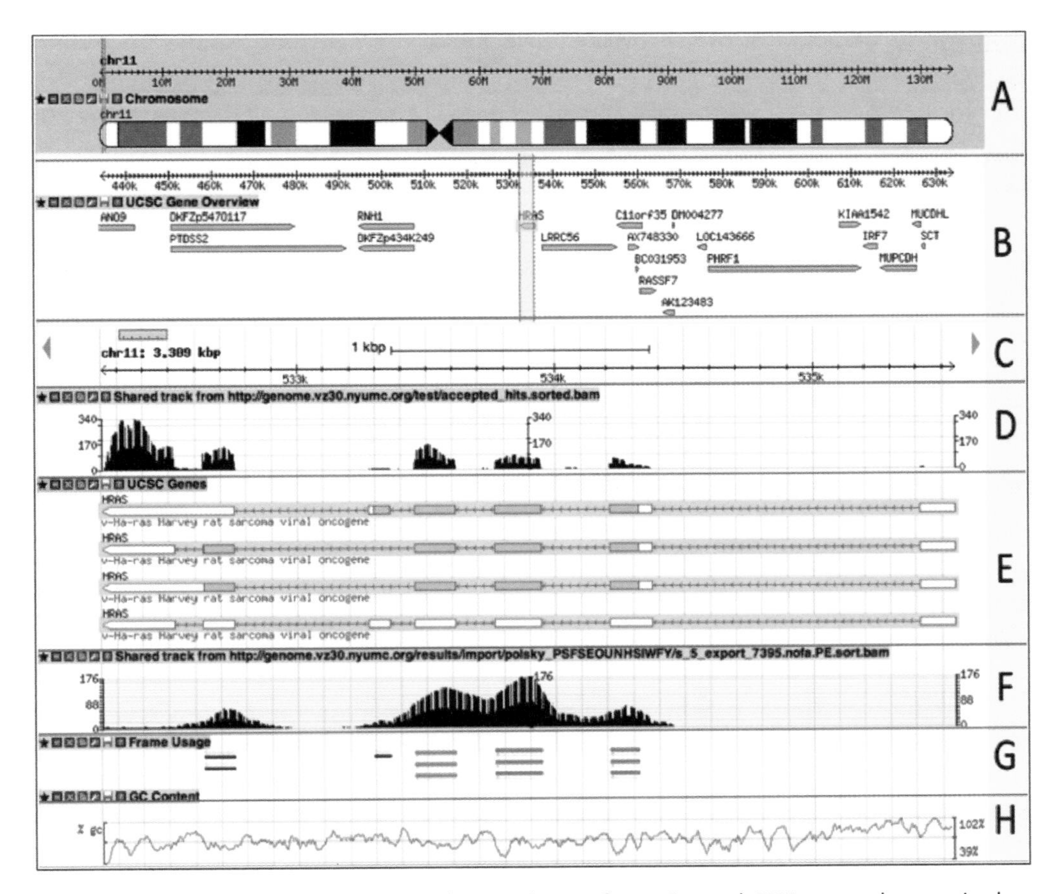

FIGURE 8. A local instance of GBrowse showing lanes of experimental RNA-seq and genomic data. (*A*) Birds-eye view of chromosome 11 with region of interest indicated by two vertical red lines. (*B*) A zoomed-in view of the gene *HRAS* (highlighted in yellow) between coordinates 530,000 and 540,000 on chromosome 11, with neighboring genes from UCSC. (*C*) The finer zoomed-in view of the *HRAS* gene region with 1-kb scale and options to scroll left and right. (*D*) A shared track (loaded by URL from a network storage device) displaying Illumina RNA-seq reads as TopHat/Cufflinks assembled transcripts in SAM/BAM format. Peaks are clearly visible aligning with the exon regions of the transcripts below. (*E*) The gene *HRAS* and its different transcript isoforms with defined gene structures (exons are in box shape, CDS are blue, whereas UTRs are white). Small arrows in intron regions and the skewed box to the left indicate the direction of transcription. (*F*) Shared Illumina Exome sequencing data in SAM/BAM format, showing the depth of coverage across annotated exons. (*G*) CDS reading frame in "musical staff" notation. (*H*) %GC content of the current view region.

GBrowse is written in Perl and is highly customizable according to the user's specific needs. The latest version of GBrowse enables users to upload custom tracks and annotations without the involvement of the maintainer/system administrator. The option of uploading files through remote URLs allows users to visualize data stored on large-scale devices located in data centers at their institution, or elsewhere on the Internet, without the need to download the data files to their desktop computer.

A local installation of GBrowse can facilitate the distribution and analysis of sequence information at a research institution. Scientists can access sequence data files located on a network storage server via a web browser in the GBrowse visualization (see Fig. 8). The following workflow has been implemented at the sequencing core facility at NYULMC.

1. Customer's sample is sequenced using any NGS technology at the core laboratory.
2. Automated base calling and demultiplexing using manufacturer-supplied software are performed.
3. After necessary quality-control steps, the reads are either aligned to the reference genome (ELAND or BWA) or a de novo assembly is performed.
4. The data are then converted to one of the formats compatible with GBrowse, preferably GFF3.
5. The data are stored on the HPC storage cluster, and the location of the GBrowse formatted file (e.g., GFF3) is uploaded as a custom track remotely.
6. The link is then shared with the investigator for further examination and analysis.
7. The custom tracks can be made permanent as part of the reference GBrowse (hg19, hg18, mm9, mm8) or hosted as a password-protected browser.

REFERENCES

Darling AE, Mau B, Perna NT. 2010. progressiveMauve: Multiple genome alignment with gene gain, loss, and rearrangement. *PLoS ONE* **5:** e11147. doi: 10.1371/journal.pone.0011147.

Rizzo JM, Buck MJ. 2012. Key principles and clinical applications of "next-generation" DNA sequencing. *Cancer Prev Res* **5:** 887–900.

Robinson JT, Thorvaldsdóttir H, Winckler W, Guttman M, Lander ES, Getz G, Mesirov JP. 2011. Integrative genomics viewer. *Nat Biotechnol* **29:** 24–26.

Staden R, Beal K, Bonfield JK. 1998. The Staden package, 1998. *Methods Mol Biol* **132:** 115–130.

Stein LD, Mungall C, Shu S, Caudy M, Mangone M, Day A, Nickerson E, Stajich JE, Harris TW, Arva A, Lewis S. 2002. The Generic Genome Browser: A building block for a model organism system database. *Genome Res* **12:** 1599–1610.

WWW RESOURCES

http://flybase.org FlyBase, a database of *Drosophila* genes and genomes.

http://genome.ucsc.edu/goldenPath/help/wiggle.html Wiggle Track Format home page, University of California, Santa Cruz, Genome Bioinformatics.

http://gmod.org/wiki/GFF3 Generic Feature Format (GFF), from Generic Model Organism Database (GMOD), is a standard file format for storing genomic features in text files.

http://qed.princeton.edu/main/Wiggle_(BED_files) Wiggle (BED) files used to upload quantitative data to GBrowse.

http://samtools.sourceforge.net/SAM1.pdf The SAM Format Specification.

http://samtools.sourceforge.net/ The SAM (Sequence Alignment/Map) format is a generic format for storing large nucleotide sequence alignments at sourceforge.net (the place to download and develop free open source software).

http://useq.sourceforge.net/useqArchiveFormat.html The USeq Compressed Binary Data Format 1.0 archive is a zip-compressed "directory" containing genomic data split by chromosome, strand, and slices of observations.

http://wormbase.org WormBase home page. Explores worm biology facilitating insights into nematode biology.

http://www.arabidopsis.org The *Arabidopsis* Information Resource (TAIR) maintains a database of genetic and molecular biology data for the model higher plant *Arabidopsis thaliana.*

http://www.genomeview.org GenomeView (T Abeel, 2008–2011) is a next-generation standalone genome browser and editor initiated in the Bioinformatics & Systems Biology (BSB) group at VIB and currently developed at the Broad Institute.

http://www.sequenceontology.org/gff3.shtml L Stein, 2010. Generic Feature Format Version 3 (GFF3), Sequence Ontology Project.

4

DNA Sequence Alignment

Efstratios Efstathiadis

Advances in sequencing technologies and dropping costs have enabled full genome sequencing at a massive scale, initiated many new projects that were not feasible with preexisting technologies, and led to a better understanding of human variation and investigation of personalized medicine (see Chapter 1). The short-length reads produced by **next-generation sequencing** (NGS) instruments, the **Poisson distribution** of the reads that results in uneven coverage of the sequenced genome, **repetitive** and hard-to-sequence regions of the genome, insertions, deletions, and machine errors that result in miscalled bases (typically on the order of 1%) require multiple genome coverage (typically 30×), resulting in millions of **short reads** per sequencing run. The **alignment** of short reads to a **reference genome** is a critical task to most NGS projects such as identifying **sequence variants** of individuals or measuring gene expression by RNA sequencing. The alignment of millions of short nucleotide reads (ranging from a few tens of to a few hundred nucleotides) to a large reference genome is a computationally intensive, time-consuming, and critical step for many NGS projects.

A variety of **alignment algorithms** and tools have been developed to address these challenges. Several **algorithms** have benefited from advances in computer processor power, computer memory size and access speed, and network interconnects (see Chapter 12). Scalable algorithms can take advantage of multithreading by using several processing cores and multiprocessing leveraging MPI (Message Passing Interface) libraries and fast interconnects between compute nodes. Algorithms have been designed to take advantage of hardware accelerators, such as field programmable gate arrays (FPGAs) and graphics processing units (GPUs), whereas others benefit by simply splitting large problems (either large collections of sequencing data sets or intense computational tasks) into smaller subproblems that can be solved independently, in parallel, either as simple embarrassingly parallel tasks or using map-reduce algorithms to distribute the load over several compute nodes. Many alignment tasks exploit the cost benefits and ready-to-use services provided by compute cloud deployments. Here we present an overview of popular alignment algorithms, following a historical path, giving emphasis to algorithms that handle the alignment task of a large collection of short nucleotide reads produced by next-generation sequencing instruments. We do not intend to provide a complete, in-depth review of each algorithm presented, but rather an introduction so that the reader can follow up for more details with the references provided.

DYNAMIC PROGRAMMING ALGORITHMS

Although at first it may seem that aligning DNA sequences is not different from simply managing strings of characters, comparing two sequences using a brute-force approach is impractical because of the large size of the strings involved. For example, there are an enormous number of possible alignments in comparing two DNA sequences with 100 nucleotides (residues) each. Because biological sequences are subject to **indels** (insertions and deletions), affine gaps, substitutions, and so on, a scoring scheme was introduced to take into account the biological nature of the comparing sequences. Dynamic programming algorithms, such as the Needleman–Wunsch (NW) algorithm (Needleman and Wunsch 1970), can calculate the best alignment based on a scoring scheme, in $n \times m$ steps. The Needleman–Wunsch algorithm performs global alignments of amino acid or nucleotide sequences, and being a global alignment algorithm, works best when the compared sequences have comparable lengths and is preferred when comparing genes with the same function. Global **sequence alignment** algorithms attempt to align every nucleotide (residue) from either sequence, thus the algorithm searches for the maximum number of nucleotides in one sequence that can be matched on the second sequence, allowing for all possible interruptions in both. In comparing the two sequences, the algorithm introduces an iterative two-dimensional (2D) matrix (the *score matrix*) in which the nucleotides from one sequence (the *database sequence*) are arranged horizontally across the top with each nucleotide corresponding to a column, whereas nucleotides from the other one (the *query sequence*) are arranged vertically down the side, with each nucleotide corresponding to a row. Matrix cells represent all possible combinations of nucleotides from both sequences, whereas paths throughout the matrix represent all possible alignments. Initially, the first row and column of the score matrix are filled with the scores of the *residue substitution score table*, a table that provides a score for all possible residue alignments (a 4×4 table in the case of nucleotide sequences) (Table 1). A penalty is introduced for every residue-to-nothing (gap) match. In the next step in finding the best alignment, every cell in the score matrix, proceeding row by row, is assigned two parameters: (1) the maximum score for an alignment ending in that point, and (2) a pointer. The pointer may be kept on a separate matrix, called the *traceback matrix*. The optimal alignment is constructed by tracing back from the

TABLE 1. A simple residual substitution score matrix (S) for DNA nucleotides

	A	C	T	G
A	1	−1	−1	−1
C	−1	1	−1	−1
T	−1	−1	1	−1
G	−1	−1	−1	1

lower-right corner to the upper-left corner cell following the pointers of every cell along the pathway.

The NW algorithm is not appropriate for determining local regions of high similarity because it introduces gaps around such regions. The Smith–Waterman (SW) algorithm (Smith and Waterman 1981) finds the optimal *local* alignment of the two compared sequences. It is based in identifying a pair of subsequences (segments) between two long sequences with the highest degree of similarity, with respect to a scoring system. As in the case of the NW algorithm, the residuals of the two sequences are arranged along the top and side of the score matrix (H), but first, in the initialization phase, the cells on the first row and column (the edges of the matrix) are filled with 0's: $H_{i,0} = H_{0,j} = 0$ for all i, j. The rest of the cells are filled with scores using a scoring system that is based on a residual substitution score table (S) and a gap penalty (d), negative values being replaced with 0s:

$$H_{i,j} = \max\{0, H_{i-1,j-1} + S_{i,j}, H_{i-1,j} - d, H_{i,j-1} - d\}.$$

The pair of segments with the highest similarity is determined by first locating the cell with the highest score and then sequentially determining the rest of the cells with a traceback procedure ending in a cell with score equal to zero. In the following example, from Hasan and Al-Ars (2007), we consider the alignment of the sequences: $A = \text{a g g t a c}$ and $B = \text{c a g c g t t g}$ using a gap penalty of $d = 2$ and a scoring scheme of $S_{i,j} = 2$ if the two residual nucleotides match, or −1 otherwise. The resulting alignment is

$$A\text{: a g} - \text{g t}$$

$$B\text{: a g c g t.}$$

The corresponding score matrix (H) and the traceback path (in bold) are shown in Table 2.

The SW algorithm has been enhanced in several ways; one such improvement was introduced by Gotoh (1982) with the affine gap model. A gap is used to account

TABLE 2. The score table (**H**) for the alignment of the two sequences arranged along the top and the side of the table

		c	a	g	c	G	t	t	g
	0	**0**	0	0	0	0	0	0	0
a	0	0	**2**	0	0	0	0	0	0
g	0	0	0	**4**	**2**	2	0	0	2
g	0	0	0	2	3	**4**	2	0	2
t	0	0	0	0	1	2	**6**	4	2
a	0	0	2	0	0	0	4	5	3
c	0	2	0	1	2	0	2	3	4

Redrawn, with permission, from Hasan and Al-Ars 2007.
The cells in bold represent the traceback path, starting from the maximum value of 6.

for an insertion or deletion in the query sequence with respect to the database sequence. From a biological perspective, a single multibase insertion or deletion event is more likely to occur than several small indels near each other. In the affine gap model, the penalty score for the first character in a gap is called *gap_open*, and the cost for each additional character is called *gap_extension*.

Both the NW and SW algorithms are based on dynamic programming and are computationally expensive, especially when large sequences are compared requiring the creation of large matrices and calculation of scores for every matrix cell. Aligning two 100-bp-long sequences, for example, requires the creation of a 100×100 matrix and the calculation of 10^4 scores. When large sequences are aligned, such calculations take a long time and consume large amounts of computational resources. In the case of the human genome, which is $\sim 3.2 \times 10^9$ bp long, the mapping of a 100-bp read requires the calculation of 10^{12} scores to fill the elements of the scoring table. Because the calculation of each individual score requires several floating-point operations, the needed computational power to calculate a single matrix element is on the order of several teraFLOPS ("FLoating-point OPeration," 1 teraFLOP $= 10^{12}$ FLOPS). Typical NGS instrument runs produce millions of reads that will require the computational power provided by a medium-sized supercomputer or Linux cluster for the alignment process. The SW algorithm can be parallelized in two fronts: first by distributing the processing of each one of the query sequences on several individual processors that can be aligned independently and in parallel. The other front involves distributing the work required for a single alignment. This is more challenging and involves calculating the scores of each cell in the score matrix H. As shown in Figure 1, the score of each cell can only be computed after the scores of neighboring cells have been calculated first. Several efforts have reported speedups of the SW algorithm using reconfigurable computing based on FPGA chips (e.g., Storaasli et al. 2007) and graphics processing units (GPUs) (Liu et al. 2006; Manavski and Valle 2008).

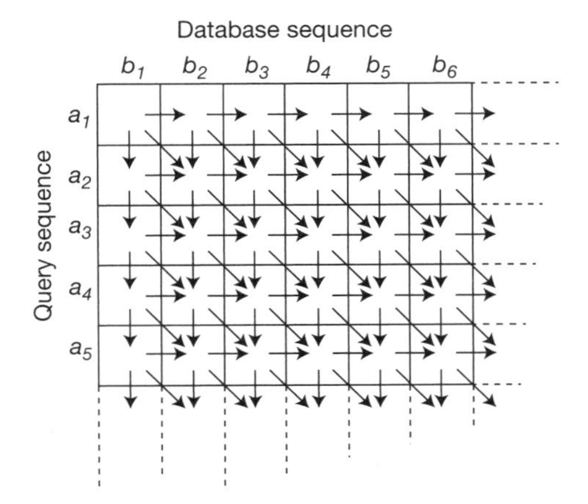

FIGURE 1. Computational dependencies in the Smith–Waterman alignment matrix. (Reproduced from Rognes and Seeberg 2000, with permission of Oxford University Press.)

Database Searching

The NW and SW algorithms are used when comparing two sequences. When a completely new sequence is discovered, however, we would like to know if it is similar to any of the already known sequences from all organisms. Sequence databases, such as **GenBank**, contain hundreds of billions of bases and are currently growing at an exponential rate, making the application of dynamic programming algorithms impractical for database searching. To improve the speed of searching, we need to resort to heuristic algorithms that make use of approximations to speed up sequence comparisons. Heuristic algorithms like FASTA and BLAST have improved the alignment running time by a large factor at the expense of sensitivity.

The typical approach of approximate methods is to determine all possible residue words of length k in the query sequence, called ***k*-tuples**, and discover all sequences in a database that contain a large number of matching k-tuples. FASTA (the name comes from FAST Alignment) was developed by Lipman and Pearson (1985) and was further improved in 1988 (Pearson and Lipman 1988). The first step in FASTA is to identify all exact matching k-tuples in both sequences. For DNA sequences, the default value of k is 6. The value of k controls the trade-off between speed and sensitivity. To expedite the process, a lookup array (hash) is built with all possible k-tuples. First, sequence A is scanned and the location of each found k-tuple is recorded in the lookup array. Then sequence B is scanned, and using the lookup array, all common k-tuples between sequences A and B are found (called *hits* or *hot spots*). The identified k-tuples are located on diagonals of the matrix with possibly several k-tuples being on the same diagonal. What is important in the algorithm is not the exact words (hits) encountered, but rather segments containing several hits (also called diagonal runs). Identified segments containing hits and spaces are scored, with k-tuples getting a positive score and spaces getting a negative score. The longer the spaces in between k-tuples (hits) in a segment, the lower will be the score for that segment. Once all the common k-tuples between the two sequences are identified, located on the matrix, combined into diagonal runs, and evaluated based on the number of common k-tuples found and the distance between matches, then the 10 best (highest scoring) diagonals are selected. These are the initial regions. The highest score of the segments is called $init_1$. The initial segments are then joined together, combining segments from close diagonals into new high-scoring local alignments, containing insertions and deletions. The highest-scoring alignment in this stage is called $init_n$. Finally, dynamic programming alignment is used over a narrow diagonal band centered on the segment with $init_1$ score to find the highest-scoring alignment, with a score marked as *opt*. For the final group of alignments, a full **Smith–Waterman alignment** is performed. FASTA reports the top-scoring alignment for each sequence encountered in the database.

The FASTA software can be downloaded via ftp from http://fasta.bioch.virginia.edu. Several sites offer web-based FASTA searches. FASTA Search Examples are available through the FASTA service at EMBL-EBI (http://www.ebi.ac.uk/Tools/sss/fasta/nucleotide.html), the KEGG service in Japan (http://www.genome.jp/tools/fasta/), or the University of Virginia server (http://fasta.bioch.virginia.edu/fasta_www2/fasta_www.cgi).

BLAST

BLAST (Basic Local Alignment Search Tool) (Gish et al. 1990) is a heuristic (approximate) algorithm that was introduced to improve the speed of database searching of FASTA by restricting searches in fewer and smaller bands around matrix diagonals. BLAST creates a hash table using as keys all possible subsequences (words) of length w (the default value for w for nucleotide sequences is 11) from the query sequence. It searches all the database sequences for exact matches to words, called *seeds*, and tries to extend each match in both directions, first without allowing gaps or mismatches. Subsequently the extended matches can be joined within restricted regions, allowing for gaps and mismatches, with an alignment score that is greater than a selected minimum, cutoff score T. As the cutoff score T increases, fewer matches are considered and the program runs faster. The restricted regions where the joining of the extended matches can proceed look like bands around diagonals on the substitution matrix or dot plot. The region restriction, rather than considering the entire matrix, is what gives BLAST its speed. However, BLAST will miss any matches outside of the diagonal band.

BLAST is a collection of algorithms. The choice of algorithm depends on the query sequence type, the purpose of the search, and the target database. NCBI provides a BLAST program selection guide, available at http://blast.ncbi.nlm.nih.gov/Blast.cgi?CMD=Web&PAGE_TYPE=BlastDocs&DOC_TYPE=ProgSelectionGuide. For nucleotide queries, BLASTN compares a nucleotide sequence with a nucleotide database, and by using the default short word size of 11, it is more sensitive that MEGABLAST. MEGABLAST is designed to find long alignments (uses a default length of 28) between similar sequences and thus is the recommended tool in the BLAST family for finding database sequences that are very similar or even identical to the query sequence. A list of available algorithms is available on the NCBI BLAST web pages: http://blast.ncbi.nlm.nih.gov/Blast.cgi. This web page provides an interface where various databases can be searched using BLAST algorithms.

Alignment of NGS Reads

DNA reads produced by NGS instruments have different characteristics from those produced by conventional **Sanger sequencing** (see Chapter 1). Individual NGS read lengths range from a few tens of to several hundred base pairs and present different error characteristics. Millions of reads are produced by a single instrument run that must be aligned to a long reference genome of the same species before application-specific software can be used. Aligning a large number of short reads to a reference requires algorithms that are fast to deal with the large amount of reads and sensitive to deal with the error characteristics of small read size. Traditional alignment techniques are computationally expensive (in terms of actual processing and memory usage), especially when large sequences are compared. Methods such as Smith–Waterman, BLAST, and FASTA use computationally demanding algorithms that are not optimized to deal with the volume of NGS data and the required sensitivity.

To address the challenges of aligning NGS reads, existing methods such as hash tables and seed-and-extend methods have been optimized, and new ones, like Burrows–Wheeler, based on suffix trees have been introduced. A large number of software tools have been developed recently to deal with the volume and characteristics of NGS reads (Li and Homer 2010). The seeding method used with BLAST is based on finding short, exact matches of seeds, called *hits*. This approach has been extended to allow for spaced or nonconsecutive seeds: seeds that include spaces in the original subsequence (Ma et al. 2002). In this approach, in addition to the length k of the original query subsequence, a template (model) with 0s and 1s is introduced specifying the relative position of the k residues in the seed. In the template, 1s represent required matches, whereas 0s represent indels or mismatches. The number of 1s is called the *weight* of the seed. For example, for a weight of $k = 6$ and using the template 1110111, the following two sequences represent seed matches: actgact and actaact. In BLAST's consecutive seed model, the template consists of 11 1s and no 0s.

Several alignment tools have been developed based on hash tables. Some tools index the **sequence reads**, whereas others index the reference genome. The Short Oligonucleotide Alignment Program (SOAP) (Li et al. 2008a) does gapped and ungapped alignment of a large number of short reads to a reference genome, using seed and hash lookup tables. It loads the reference genome in memory and creates seed index tables for all possible sequences. The difference between SOAP and other tools like MAQ (Li et al. 2008b) and SeqMap (Jiang and Wong 2008) is that the latter ones load the query read sequences into memory and create seed index tables for the reads instead of the reference genome. A filtering process that helps reduce the number of potential matches, implemented on the above tools, is to split each read into several fragments and then find exact matches to only some of the fragments. For example, to accept two mismatches, the read is split into four fragments with at most two of the four fragments containing a mismatch. We can then find hits with two mismatches by trying all six combinations of the four fragments as seeds. An improved version of SOAP, SOAP2 (Li et al. 2009), uses the **Burrows–Wheeler transformation** (BWT) compressed index, instead of the seed-based algorithm, to index the reference genome in memory. It is an order of magnitude faster, significantly reduces the memory footprint, and supports a wide range of sequence read lengths.

ELAND (Efficient Large-scale Alignment of Nucleotide Databases) is Illumina's aligner used with sequence reads produced by its instruments. It is part of the CASAVA (Consensus Assessment of Sequence and Variation) bundle that performs alignment of reads to a reference genome, subsequent variant analysis, and read counting (Ilumina 2011a,b). ELAND was one of the first aligners to introduce spaced seeds in short read alignment. Similarly to MAQ, it creates a hash table and index for the sequence reads rather than the reference genome. An updated version, ELANDv2, introduced multiseed and gapped alignments. Gapped alignments handle insertions and deletions by extending each candidate alignment to the full length of the read. Multiseed alignments align the first seed of 32 bases and consecutive seeds separately. Every candidate alignment gets a probability score (*P*-value) based on base quality and position of mismatches. The *P*-value of the candidates determines the alignment score.

Burrows–Wheeler Transformation (BWT)–Based Aligners

A new family of aligners (Bowtie, BWA, SOAP2) uses the Burrows–Wheeler transformation (BWT) (Burrows and Wheeler 1994) of the reference genome to increase the speed and reduce the memory footprint of the alignment process. The *trie* (the name comes from the word re*trie*val) for a reference string is a data structure that allows fast string operations, such as string matching. A *trie* contains all possible substrings (prefixes for a prefix *trie* or suffixes for a suffix *trie*) of a reference string. Each string character is stored on a node (or edge) with each substring being delimited with a special character (usually $ marks the end of substrings in suffix *tries*, whereas ^ marks the start of a substring for prefix *tries*) (see Fig. 2). In a prefix *trie*, concatenating all characters on the nodes from a leaf to the root forms a unique substring. Because exact repeats of the reference string are located under the same path of a *trie*,

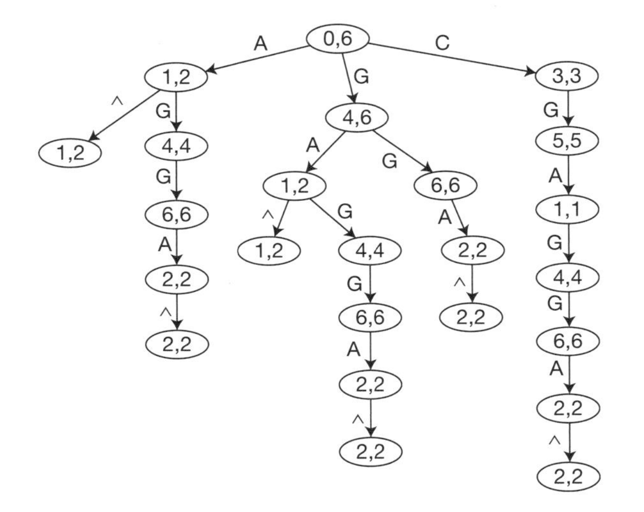

FIGURE 2. A prefix *trie* of string AGGAGC. Starting from a leaf node and concatenating the characters on all the nodes to the root forms a unique substring. The character ^ marks the start of the substring. The numbers in a node indicate the suffix array interval for that node. (Reproduced from Li and Homer 2010, with permission of Oxford University Press.)

the query string needs to be matched (aligned) to only one copy of the repetitive region. This is the main reason algorithms based on *tries* and their derivative data structures are fast. The time required to test if a query string is an exact match of a reference string represented by a *trie* is linear in the length of the query string. Although the *trie* approach is fast, it takes up a lot of memory space. The space required is on the order of the square of the length of the reference string (Li and Homer 2010).

Finding every occurrence of a substring in a large reference string is equivalent to finding all suffixes that begin with that substring. Once all suffixes have been ordered lexicographically, they can be searched efficiently with a binary search. This simplistic method is shown in the following example, reproduced from (http://plindenbaum.blogspot.com/2010/01/elementary-school-for-bioinformatics.html).

Starting from a *reference string T* = gggtaaagctataactattgatcaggcgtt, we can first list all suffixes:

[00] gggtaaagctataactattgatcaggcgtt
[01] ggtaaagctataactattgatcaggcgtt
[02] gtaaagctataactattgatcaggcgtt
[03] taaagctataactattgatcaggcgtt
[04] aaagctataactattgatcaggcgtt
[05] aagctataactattgatcaggcgtt
[06] agctataactattgatcaggcgtt
[07] gctataactattgatcaggcgtt
[08] ctataactattgatcaggcgtt

[09] tataactattgatcaggcgtt

[10] ataactattgatcaggcgtt

[11] taactattgatcaggcgtt

[12] aactattgatcaggcgtt

[13] actattgatcaggcgtt

[14] ctattgatcaggcgtt

[15] tattgatcaggcgtt

[16] attgatcaggcgtt

[17] ttgatcaggcgtt

[18] tgatcaggcgtt

[19] gatcaggcgtt

[20] atcaggcgtt

[21] tcaggcgtt

[22] caggcgtt

[23] aggcgtt

[24] ggcgtt

[25] gcgtt

[26] cgtt

[27] gtt

[28] tt

[29] t

All suffixes are then sorted lexicographically:

[04] aaagctataactattgatcaggcgtt

[12] aactattgatcaggcgtt

[05] aagctataactattgatcaggcgtt

[13] actattgatcaggcgtt

[06] agctataactattgatcaggcgtt

[23] aggcgtt

[10] ataactattgatcaggcgtt

[20] atcaggcgtt

[16] attgatcaggcgtt

[22] caggcgtt

[26] cgtt

[08] ctataactattgatcaggcgtt
[14] ctattgatcaggcgtt
[19] gatcaggcgtt
[25] gcgtt
[07] gctataactattgatcaggcgtt
[24] ggcgtt
[00] gggtaaagctataactattgatcaggcgtt
[01] ggtaaagctataactattgatcaggcgtt
[02] gtaaagctataactattgatcaggcgtt
[27] gtt
[29] t
[03] taaagctataactattgatcaggcgtt
[11] taactattgatcaggcgtt
[09] tataactattgatcaggcgtt
[15] tattgatcaggcgtt
[21] tcaggcgtt
[18] tgatcaggcgtt
[28] tt
[17] ttgatcaggcgtt

The resulting structure (sorted suffixes) can be searched efficiently for a *query* substring (such as aagctataacta) using *binary search*. The suffix array uses an index containing the location of each suffix in the original string, and thus it provides the location of the query substring in the reference string. This method of transforming a long string (such as a reference genome) into a sorted list can be easily parallelized because each query read can be searched independently. However, it is inefficient in terms of memory space usage. Assuming that an integer takes up 4 bytes of memory space, then a suffix array for the human genome's 3.2×10^9 integers will require 12 GB.

An efficient way to search for short sequence reads in a large reference genome using a small memory footprint is to use the FM index (Ferragina and Manzini 2000) on the BWT of the reference (see Fig. 3). To generate the BWT, we add the end-of-string character (usually $) at the end of *T*, which is lexicographically less than all characters in *T*. Generating all cyclic rotations of the text *T$* results in as many strings as the number of characters in the original string plus one, with the rotated strings forming the rows of the so-called *Burrows–Wheeler matrix*. The rows are then sorted lexicographically, and the resulting last column (the rightmost

T$ = c t g a a a c t g g t $

c	t	g	a	a	a	c	t	g	g	t	$
t	g	a	a	a	c	t	g	g	t	$	c
g	a	a	a	c	t	g	g	t	$	c	t
a	a	a	c	t	g	g	t	$	c	t	g
a	a	c	t	g	g	t	$	c	t	g	a
a	c	t	g	g	t	$	c	t	g	a	a
c	t	g	g	t	$	c	t	g	a	a	a
t	g	g	t	$	c	t	g	a	a	a	c
g	g	t	$	c	t	g	a	a	a	c	t
g	t	$	c	t	g	a	a	a	c	t	g
t	$	c	t	g	a	a	a	c	t	g	g
$	c	t	g	a	a	a	c	t	g	g	t

												S
$	c	t	g	a	a	a	c	t	g	g	t	11
a	a	a	c	t	g	g	t	$	c	t	g	3
a	a	c	t	g	g	t	$	c	t	g	a	4
a	c	t	g	g	t	$	c	t	g	a	a	5
c	t	g	a	a	a	c	t	g	g	t	$	0
c	t	g	g	t	$	c	t	g	a	a	a	6
g	a	a	a	c	t	g	g	t	$	c	t	2
g	g	t	$	c	t	g	a	a	a	c	t	8
g	t	$	c	t	g	a	a	a	c	t	g	9
t	$	c	t	g	a	a	a	c	t	g	g	10
t	g	a	a	a	c	t	g	g	t	$	c	1
t	g	g	t	$	c	t	g	a	a	a	c	7

FIGURE 3. Starting from the original string *T* = ctgaaactggt, we introduced the end-of-string symbol $ and created all cyclic permutations, generating the Burrows–Wheeler matrix (*top* matrix). We then sorted the rows lexicographically. The rightmost column of the resulted matrix is the Burrows–Wheeler transformation of *T*, BWT(*T*) = tgaa$attggcc. *S* is the suffix array indicating the position of each suffix in the original string *T*.

column) is the BWT(*T*), of the original string. The first column serves as a *genome dictionary*. For example, starting from *T* = ctgaaactggt (http://vimeo.com/channels/202195/22845446), we first create all cyclic permutations of *T$* and then sort them lexicographically. The last column of the resulted matrix is the BWT transformation of *T*, tgaa$attggcc. The *suffix array S* consists of integers indicating the location of each character in the original string *T*. Basically, it provides the position of the suffixes (hence it is called the suffix array) in lexicographical order.

The suffix array and the BWT are related as follows:

$$\mathrm{BMT}[i] = \$ \quad \text{if } S[i] = 0,$$

$$\mathrm{BMT}[i] = T[S[i] - 1] \quad \text{otherwise.}$$

This property allows us to construct the BWT from the suffix array in linear space and time.

The BWT transformation of a text has runs of repeated character, making it more compressible. As a matter of fact, the transformation is being used in data compression tools like *bzip2*. BWT is a lossless transformation (the original string of characters can be reproduced because the transformation simply changes the order of the characters). This is due to the *property* of last-first (LF) mapping of the matrix of sorted cyclic rotations (also called the *unpermute* or *walk-left* algorithm): The i-th occurrence of a character c in the last column corresponds to the same character as the i-th occurrence of c in the first column (Langmead et al. 2009). This property can reverse the transformation and produce the original reference string T (see Fig. 4).

Repeated use of the LF mapping is also applied in *exact mapping*. Because the rows are sorted, exact matches of the query string, if any of them exist, will appear in a range of consecutive rows. Thus, searching for a sequence alignment in a reference string is equivalent to finding the interval in the suffix array for the substrings that match the query. Figure 5 shows the exact matching of the query string Q = ctgg in the reference T = ctgaaactggt (http://vimeo.com/channels/202195/22845446). We start by locating the range of rows in the first column of the matrix (the *dictionary* column) that include `g`, the rightmost character in the query string Q. We mark the top row that includes `g` as the `top` (row 6) and the bottom row as the `bot` (row 8). For the next character in Q (another `g`) moving from right to left, we map the top and bottom rows to the rows whose first character corresponds to the row's last character, as if that character were a `g`. The process repeats for the rest of the characters in the query string Q. Using the values of the suffix array elements in the determined interval (range of rows), we can find the positions of the exact matches in the original string. If there is an exact match, then a single suffix array interval is found. Once we know that there is an exact match, we need to find the reference position on string T. One way to find the location is to look up the suffix array element, as, for example, in Figure 4, the SA element 5 points to location 6 on the original string. But this approach requires storing the entire suffix array in memory and then constructing the BWT from the suffix array. The memory requirements for storing the suffix array entirely in memory can be very large and can reach several gigabytes, restricting the size of the reference strings that can be processed. In principle, an array of n integers requires $4n$ bytes, whereas the size of the original T and the BWT(T) requires n bytes each. In case of the small DNA alphabet, we require 2 bits for each base, thus $2n$ bits for each string [T and BWT(T)].

One efficient way to reduce the memory footprint restriction is to construct the suffix array in pieces, a block at a time, then compute the corresponding BTW and discard the SA block (Karkkainen 2007).

A different approach from using the suffix array is to "walk-left" to the beginning of the text and, counting the number of steps that are required to reach the beginning of the reference string, this will point to the exact reference location of the query string in T. This is a slow process. The Ferragina and Manzini (2000) algorithm that is used in Bowtie to search BWT-transformed reference strings with a small

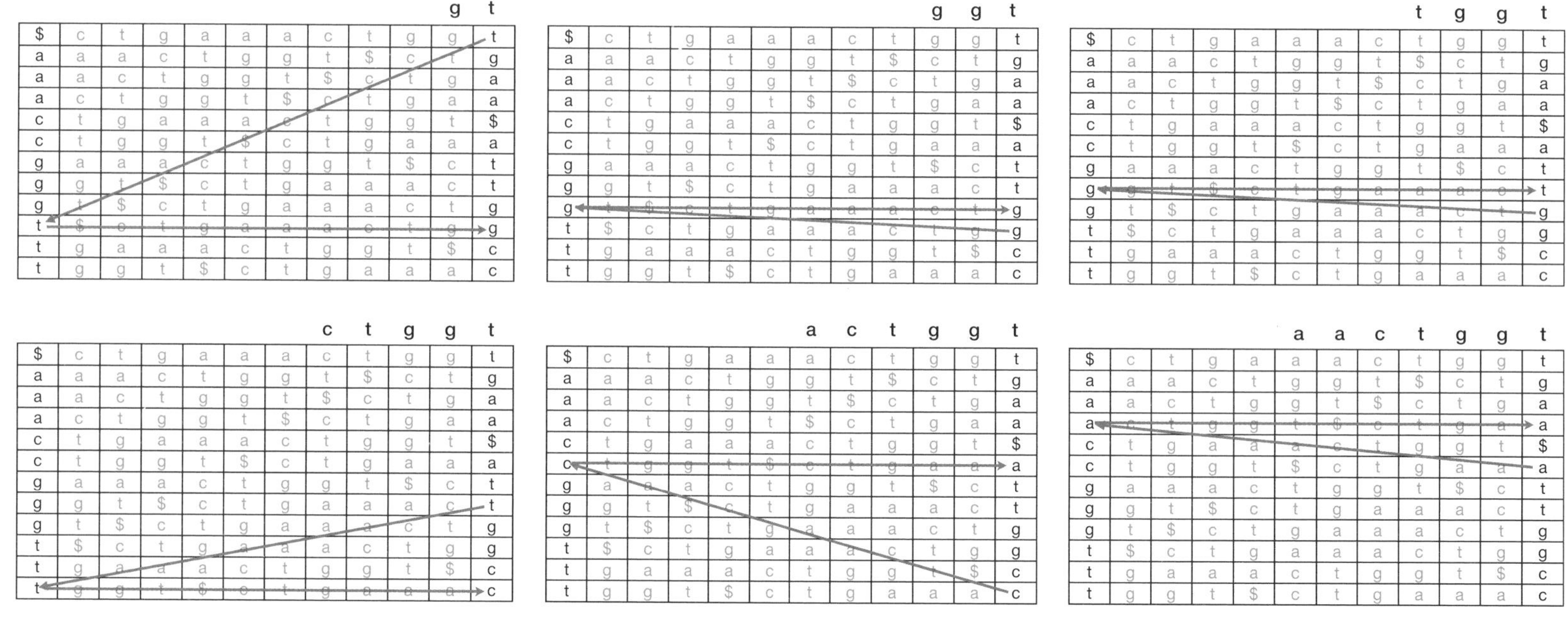

FIGURE 4. Repeated use of the last-first (LF) mapping reproduces the original string *T*. The same method is used in exact mapping.

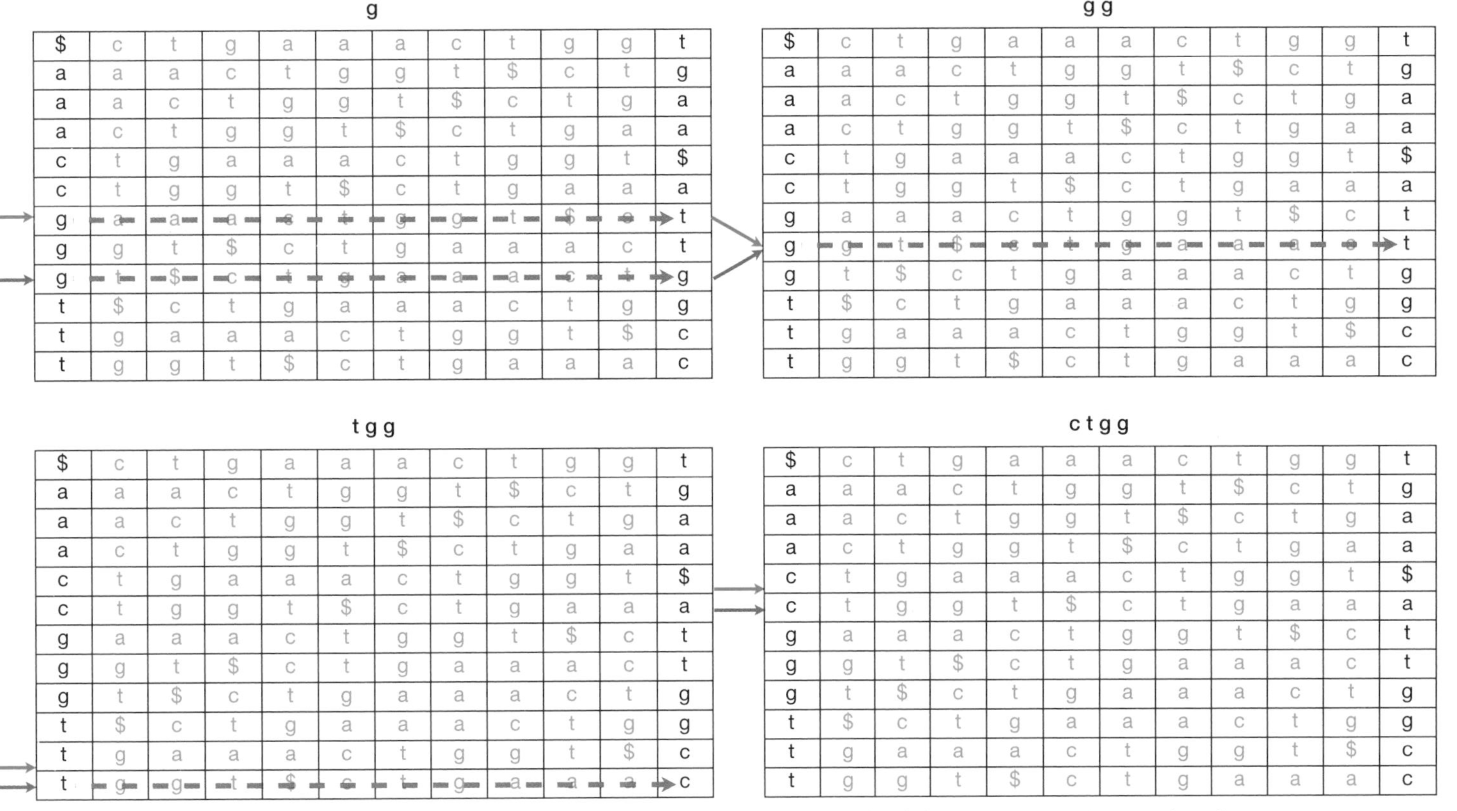

FIGURE 5. Steps to determine the suffix array interval for the exact match of the query string Q = ctgg in reference T = ctgaaactggt. There is only one location in the reference where the query mapped on the reference $S(5) = 6$.

memory footprint is a hybrid solution of the two solutions mentioned above: using samples of the suffix array and "walk-left" to the next sampled row.

Bowtie (Langmead et al. 2009) introduces a "backtracking" mechanism to account for mismatches. When exact match returns an empty range, the tool backtracks to a position with minimum quality and substitutes a different base. Bowtie's default policy is to allow for up to two mismatches in the high-quality region of the sequence read (first 28 characters). Bowtie reports a 30-fold increase in speed over hash-based algorithms with small loss of sensitivity.

SOAP2 (Li et al. 2009), an updated version of SOAP (Li et al. 2008a), is based on the BWT algorithm. It constructs a hash table to accelerate searching for the location of a read in the BWT reference index. The reference index is partitioned into a large number of blocks, and very few search interactions are sufficient to identify the exact location inside a block. SOAPaligner, part of the SOAP2 package, reports an order-of-magnitude increase in alignment speed and a large reduction in memory requirements when compared with the original SOAP hash-based algorithm. One million single-end reads (35-bp-long reads) can be aligned to the human reference genome in 2 min.

REFERENCES

Burrows M, Wheeler DJ. 1994. *A block sorting lossless data compression algorithm.* Systems Research Center Technical Report No 124. Digital Equipment Corporation, Palo Alto, CA.

Ferragina P, Manzini G. 2000. Opportunistic data structures with applications. In *Proceedings of the 41st Symposium on Foundations of Computer Science* (FOCS 2000), pp. 390–398. IEEE Computer Society Press, Piscataway, NJ.

Gish W, Miller W, Myers EW, Altschul SF, Lipman DJ. 1990. A basic local alignment search tool. *J Mol Biol* **215:** 403–410.

Gotoh O. 1982. An improved algorithm for matching biological sequences. *J Mol Biol* **162:** 705–708.

Hasan L, Al-Ars Z. 2007. Performance improvement of the Smith–Waterman Algorithm. In *Proceedings of the 18th Annual Workshop on Circuits, Systems and Signal Processing (ProRISC 2007),* November 29–30, 2007, Veldhoven, The Netherlands.

Illumina, Inc. 2011a. *CASAVA v1.8 User Guide.* Illumina Proprietary, Part# 15011196 Rev B, May 2011. http://biowulf.nih.gov/apps/CASAVA_UG_15011196B.pdf.

Illumina, Inc. 2011b. *Improved accuracy for ELAND and variant calling.* Illumina Technical Note. Pub. No. 770-2011-005. http://www.illumina.com/documents/products/technotes/technote_eland_variantcalling_improvements.pdf.

Jiang H, Wong WH. 2008. SeqMap: Mapping massive amount of oligonucleotides to the genome. *Bioinformatics* **24:** 2395–2396.

Karkkainen J. 2007. Fast BWT in small space by blockwise suffix sorting. *J Theor Comput Sci* **387:** 249–257.

Langmead B, Trapnell C, Pop M, Salzberg SL. 2009. Ultrafast and memory-efficient alignment of short DNA sequences to the human genome. *Genome Biol* **10:** R25. doi: 10.1186/gb-2009-10-3-r25.

Li H, Homer N. 2010. A survey of sequence alignment algorithms for next generation sequencing. *Brief Bioinform* **11:** 473–483.

Li R, Li Y, Kristiansen K, Wang J. 2008a. SOAP: Short oligonucleotide alignment program. *Bioinformatics* **24:** 713–714.

Li H, Ruan J, Durbin R. 2008b. Mapping short DNA sequencing reads and calling variants using mapping quality scores. *Genome Res* **18:** 1851–1858.

Li R, Yu C, Li Y, Lam TW, Yiu SM, Kristiansen K, Wang J. 2009. SOAP2: An improved ultrafast tool for short read alignment. *Bioinformatics* **25:** 1966–1967.

Lipman DJ, Pearson WR. 1985. Rapid and sensitive protein similarity searches. *Science* **227:** 1435–1441.

Liu Y, Huang W, Johnson J, Vaidya S. 2006. GPU-accelerated Smith–Waterman. *Lecture notes in computer science*, Vol. 3996, pp. 188–195. Springer-Verlag, Berlin.

Ma B, Tromp J, Li M. 2002. PatternHunter: Faster and more sensitive homology search. *Bioinformatics* **18:** 440–445.

Manavski SA, Valle G. 2008. CUDA compatible GPU cards as efficient hardware accelerators for Smith–Waterman sequence alignment. *BMC Bioinformatics* **9:** S10. doi: 10.1186/1471-2105-9-S2-S10.

Needleman SB, Wunsch CD. 1970. A general method applicable to the search for similarities in the amino acid sequence of two proteins. *J Mol Biol* **48:** 443–453.

Pearson WR, Lipman DJ. 1988. Improved tools for biological sequence comparison. *Proc Natl Acad Sci* **85:** 2444–2448.

Rognes T, Seeberg E. 2000. Six-fold speed-up of Smith–Waterman sequence database searches using parallel processing of common microprocessors. *Bioinformatics* **16:** 699–706.

Smith TF, Waterman MS. 1981. Identification of common molecular subsequences. *J Mol Biol* **147:** 195–197.

Storaasli O, Yu W, Strenski D, Malby J. 2007. *Performance evaluation of FPGA-based biological applications.* Cray Users Group Proceedings, May 2007, Seattle WA.

WWW RESOURCES

http://blast.ncbi.nlm.nih.gov/Blast.cgi?CMD=Web&PAGE_TYPE=BlastDocs&DOC_TYPE=ProgSelectionGuide Basic Local Alignment Search Tool (BLAST) home page, U.S. National Library of Medicine, Bethesda, MD.

http://fasta.bioch.virginia.edu FASTA Sequence Comparison, University of Virginia, Charlottesville.

http://fasta.bioch.virginia.edu/fasta_www2/fasta_www.cgi Search databases with FASTA. © W.R. Pearson, University of Virginia, Charlottesville.

http://plindenbaum.blogspot.com/2010/01/elementary-school-for-bioinformatics.html P. Lindenbaum, 2010. Elementary School for Bioinformatics: Suffix arrays. A blog regarding bioinformatics, semantic web, comics, and social networks.

http://vimeo.com/channels/202195/22845446 D. Wall and P. Tonellato, 2011. Harvard Medical School BMI 714 Introduction to Next-Generation Sequencing: Methods, Analysis, and Applications. Video blog.

http://www.ebi.ac.uk/Tools/sss/fasta/nucleotide.html FASTA nucleotide similarity search. European Bioinformatics Institute, European Molecular Biology.

http://www.genome.jp/tools/fasta/ FASTA KEGG search, Kyoto, Japan.

5

Genome Assembly Using Generalized de Bruijn Digraphs

D. Frank Hsu

The sequencing of new, unknown genomes (**de novo sequencing**) is one of the key applications of **next-generation sequencing** (NGS) technologies. This chapter discusses recent results using the interconnection properties of de Bruijn–based digraphs (directed graphs) to assemble NGS short reads into complete genome sequences. First, the concept of **de novo assembly** is introduced and a general model is provided. The early **DNA fragment assembly algorithms** initiated in the 1980s are reviewed. Then the concept of a **de Bruijn digraph** is introduced, and various properties of its interconnection are explored. More recent results on de novo assembly using short sequence reads are reviewed. Two classes of digraph—the de Bruijn–based digraph and the generalized de Bruijn digraph—are defined based on the de Bruijn interconnection. The existence of an Eulerian path in each of these generalized de Bruijn digraphs gives rise to a DNA sequence of the original genome. This approach produces improved genome assembly results with a greatly reduced computational effort for large sets of sequence fragments. Motivated by the work of Flicek and Birney (2009) and others, we describe a general genome assembly framework using de Bruijn–based and generalized de Bruijn digraphs.

DNA FRAGMENT ASSEMBLY

The de novo sequencing of bacterial and other microbial genomes has become a common, almost routine laboratory activity, but de novo sequencing of large eukaryotic genomes remains challenging because of large genome size and the high content of **repetitive DNA**. All NGS de novo sequencing projects rely on **shotgun sequencing** approaches, which generate large sets of short sequence reads. Many copies of the DNA from the target genome are randomly fragmented, then these fragments are sequenced and the original DNA sequence is reconstructed (*assembled*) using computational algorithms and computer programs (Staden 1980).

To consider the shotgun assembly problem, suppose that the original DNA sequence is q bp long, that is, a string S with individual bases at positions s_1, s_2, ..., s_q. A DNA sequencing method will create a large number (m) of randomly generated subfragments (reads) of S: $f_1, f_2, \ldots, f_m$ of length n, where n is much smaller than q. Some NGS methods create sequence fragments that are all of the same length (i.e., Illumina HiSeq), whereas others create fragments with a range of different lengths (i.e., 454). Because the sequence is randomly sampled, the average depth of **coverage** (c) is roughly $c = (mn)/q$. If the locations of the fragments are uniformly distributed along the sequence S, then the depth of coverage at a specific base s_x follows a **Poisson random probability distribution** with mean c (Idury and Waterman 1995).

The fragment assembly algorithms used by the original Staden assembly package (Staden 1980) followed the Overlap-Layout-Consensus (OLC) model. First, all pairs of fragments were examined for significant "Overlap." Then an approximate "Layout" or tiling of the fragments was established. Third, **multiple alignments** were made of all overlapping fragments, and a "**consensus**" **sequence** was created from the multiple alignments. The OLC model works well for the assembly of small regions of DNA with long reads, but it does not scale well to the assembly of entire genomes with short reads. If the overlap (k bases) required to join two fragments is small, then many such overlaps will be found for each fragment, leading to many incorrect assemblies. If the required overlap is large, then few such overlaps will be found, and many fragments that overlap by less than the required amount will not be assembled; thus the overall coverage of the genome must be very deep in order to achieve a complete assembly. The initial phase of searching for overlaps is computationally challenging for a large DNA sequence (i.e., an entire genome) because every fragment must be tested against both ends of every other fragment, leading to $2\ (m^2)$ comparisons. Assembly of real genomes is further complicated by the presence of repeated sequences, which may be longer than the read length and therefore larger than any possible overlap size. For large eukaryotic genomes, there are repeated regions so large that complete de novo assembly is impossible with short reads. Another challenge is created by sequencing errors. Current NGS technologies produce errors at the rate of 0.5%–1%; thus a pair of fragments with a true overlap may contain one or more mismatching bases in the overlap region. If the assembly algorithm allows for mismatches, then false overlaps will also be found between fragments from nearly identical regions of the genome.

The OLC model to assemble DNA fragments is similar to the concept of a jigsaw puzzle where one wants to reconstruct the original puzzle (sequence) by examining the fit between many pairs of pieces (subsequences). The differences can be summarized in two characteristics. First, the jigsaw is a partition and recombination problem, whereas the OLC involves fragmentation (with overlaps) and reassembly. Second, the scale involved in the OLC model can be prohibitively high. The OLC approach can be viewed as a graph, where each sequence read is a node and each overlap is an edge (or an arc) that connects two nodes. In an ideal assembly, the graph will have a single solution as a path that connects all nodes and visits each node exactly once (this is known as a *Hamiltonian path*), and then the final assembled sequence can be read from the arcs that follow the path. In a similar manner, the perfect assembly can be described as an Eulerian path that visits each edge exactly once. In fact, the OLC problem is NP-hard in the sense that it is unknown if a polynomial time algorithm exists to solve the problem.

Sequencing by Hybridization

In the late 1980s, a new DNA sequencing approach, called sequencing by hybridization (SBH), was proposed (see Pevzner et al. 1989; Idury and Waterman 1995). In this approach, a two-dimensional (2D) matrix (or grid) of oligonucleotide probes for all possible short sequences of a length k (***k*-tuples**) is established. This matrix of probes is often referred to as a "sequencing chip." A single-stranded DNA fragment is then applied to the sequencing chip so that it can hybridize with all of its complementary probes in the matrix. The pattern of hybridization can then be used to reconstruct the original sequence of S. The complexity of the reconstruction problem varies with different lengths of the probes. For example, if k is small, say $k = 2$, we do not get much information because all 16 possible 2-tuples are probably present in multiple locations in the original DNA sequence S. On the other hand, a great deal of information is gained if we make k as large as possible. But to construct a chip containing all possible k-tuple sequences may not be experimentally feasible. For example, with probes of length 16, there are 4^{16} k-tuples (more than 4 billion). Although SBH never became popular for practical sequencing applications, the theoretical work on k-tuples developed by Pevzner, Waterman, and others became the foundation for many high-speed assembly and **alignment algorithms**.

Each of the two methods—SBH and shotgun sequencing—has advantages and disadvantages. Idury and Waterman (1995) were able to combine the advantages of both SBH and shotgun methods and minimize their disadvantages (see Remark 1).

Remark 1: Fragment Assembly Algorithm (Idury and Waterman 1995)
Set m, n, and k as positive integers greater than or equal to 2 (≥ 2).

1. Input: fragment sequence $f_1, f_2, \ldots, f_m$ and $|f_i| = n << q$.
   a. Construct the set of k-tuples, which are substrings of each fragment.
   b. Take the Union of all k-tuples of all fragments.

c. Construct the "spectrum digraph G" on the $(k-1)$-tuples for the k-tuples from each fragment in (a).

2. Find Eulerian tours in G and infer the original sequence S.
3. Align the fragments to produce sequence S.

In the spectrum digraph G, an Eulerian path (if it exists) would correspond to a DNA sequence S so that $G(S) = G$ (Idury and Waterman 1995). A de Bruijn digraph is a good candidate to be used to study the spectrum property in the fragment assembly.

de Bruijn Digraphs

In this section, we review the mathematical structure called the de Bruijn digraph. Included are its definition in three different respects, its various properties, and its relation to de Bruijn sequence. The use of de Bruijn digraphs for DNA fragment assembly started around 1995 and became more popular for de novo DNA assembly around 2005 as high-throughput NGS technologies became widely available. The key simplifying concept of a de Bruijn graph is to create nodes for DNA ***k*-mers** that are smaller than the read length produced by NGS (substrings of each read) and to connect these nodes by arcs when the k-mers overlap by all but one base, using a computationally simple exact matching approach. Reads are mapped across a series of nodes, following the arcs in order (a "directed graph"). Because a genome has a fixed number of k-mers of a given size, the graph does not become more complex when more sequence data are obtained (deeper coverage), and the mapping of reads to the graph is a linear rather than a quadratic problem.

A directed graph $G = (V, E)$ consists of the finite set V of n nodes $\{v_1, v_2, \ldots, v_n\}$ and the finite set E of directed arcs $\{e = (v_i, v_j) \mid 0 < i,j < n + 1\}$ that connects node v_i to node v_j. (See Fig. 1 for $V = \{a, b, c\}$ and $E = \{(a, b), (b, c), (c, a)\}$.)

A k-ary number is an integer in the set $[0, k-1] = \{0, 1, 2, \ldots, k-1\}$. An n-digit k-ary number is a representation of a sequence of n k-ary numbers, $a_1a_2\cdots a_n$, where a_i is in $[0, k-1]$ for $1 \leq i \leq n$. The n-dimensional k-ary de Bruijn digraph, written as $D(k, n)$, is a directed graph with k^n nodes, each labeled as an n-digit k-ary number $a_1a_2\cdots a_n$. Each component a_i is a number in $[0, k-1]$, the set of all numbers from

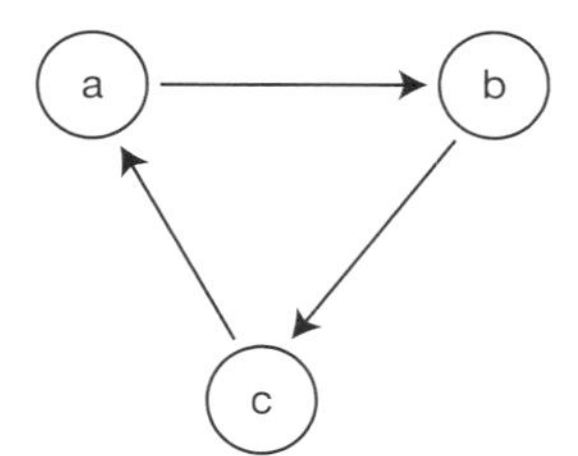

FIGURE 1. A directed graph with nodes a, b, c and arcs (a, b), (b, c), and (c, a).

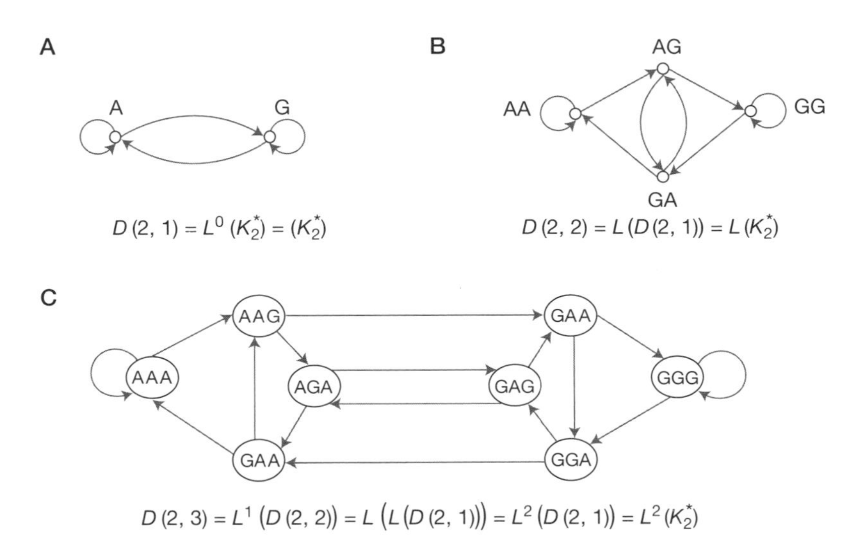

FIGURE 2. de Bruijn digraph as line digraph of $L^{i-1}(K_2^*)$ for (*A*) $D(2, 1)$ when $i = 1$, (*B*) $D(2, 2)$ when $i = 2$, and (*C*) $D(2, 3)$ when $i = 3$, respectively.

0 to $k - 1$ inclusively. Node $a_1a_2 \cdots a_n$ is adjacent to k other nodes $a_2a_3 \cdots a_n\alpha$, where α is in $[0, k - 1]$. It is easy to see that each node in $D(k, n)$ has in-degree and out-degree equal to k counting self-loops. The concept of a de Bruijn digraph was defined in the 1940s by de Bruijn (1946). An equivalent definition was defined in the 1980s by Imase and Itoh (1981) and Reddy et al. (1980) independently. In the case of a DNA sequence, the set $[0, k - 1]$, where $k = 4$, is the set of four DNA molecular components {A, C, G, T}. A de Bruijn digraph $D(k, n + 1)$ can be constructed from a de Bruijn digraph $D(k, n)$ by defining the line digraph as follows. Let G be a directed graph. The line digraph $L(G)$ is the digraph with node set $V(L(G)) = E(G) =$ arc set of G. Node (u, v) is adjacent to node (v, w) in $L(G)$ if arc (u, v) is incident to the node v and arc (v, w) is incident from the node v in G. Let K_t^* be the complete symmetric directed graph on t nodes with a self-directed loop at each node. Let $L(G)$ be the line digraph of G. Then we have $D(k, n + 1) = L^n(K_t^*)$ (Fiol et al. 1984; Grammatikakis et al. 2001, Chapter 6). The case $k = 2$ is illustrated in Figure 2 for $D(2, 1)$, $D(2, 2)$, and $D(2, 3)$, where the set of DNA molecular components is {A, G}. In the case of D(2, 3), self-loops are the arcs from AAA to AAA and GGG to GGG (Fig. 2C). The reader is referred to the book by D. West (2001) for definitions, properties, and applications of graphs and digraphs.

The de Bruijn digraph has another important property. It is closely related to a combinatorial object called the de Bruijn sequence. The k-ary de Bruijn sequence $S(k, n)$ is a circular k-ary sequence of length k^n, where each of the k^n different k-ary sequences of length n appears exactly once as a subsequence of length n.

de Bruijn sequences $S(2, 2)$, $S(2, 3)$, $S(3, 2)$, and $S(4, 2)$ and their corresponding k-ary subsequence of length n are listed as follows:

S(2, 2): {A, G}: AGGA; AG, GG, GA, AA;

S(2, 3): {A,G}: GAGGGAAA; GAG, AGG, GGG, GGA, GAA, AAA, AAG, AGA;

S(3, 2): {A, G, T}: GTTAAGGAT; GT, TT, TA, AA, AG, GG, GA, AT, TG;

S(4, 2): {A, C, G, T}: TGCCAAGGTCAGGTCG; TG, GC, CC, CA, AA, AG, GG, GT, TC, CA, AG, GG, GT, TC, CG, GT.

One observes that k^n different k-ary n-tuples in $S(k, n)$ is, in fact, the nodes in the de Bruijn digraph $D(k, n)$. For example, de Bruijn digraphs $D(2, 2)$ and $D(2, 3)$ in Figures 2B and 2C correspond to de Bruijn sequences $S(2, 2)$ and $S(2, 3)$, respectively. In this regard, the de Bruijn sequence can be viewed as the result of traversing the de Bruijn digraph in Hamiltonian order in which the traversal visits each node exactly once. Binary de Bruijn sequences $S(2, n)$ have been used since 1894 but continue to be rediscovered in a variety of contexts and applications (see Imase and Itoh 1981; Grammatikakis et al. 2001).

So far we have defined the concept of a de Bruijn digraph in several different contexts such as alphabets, line digraphs, and de Bruijn sequences. de Bruijn digraphs have a variety of important properties. They are summarized in the following (see also Grammatikakis et al. 2001).

Remark 2: Let k, n, t be positive integers greater than or equal to 2. The k-ary n-tuple de Bruijn digraph $D(k, n)$, which has k^n nodes and k^{n+1} arcs, has the following properties:

a. $D(k, n)$ consists of a Hamiltonian cycle. (A Hamiltonian cycle is a cycle that consists of each of the nodes exactly once.)
b. $D(k, n)$ consists of a cycle of length t, for any t, $2 \leq t \leq k^n$ (see, e.g., Lempel 1971).
c. $D(k, n)$ consists of an Eulerian tour for k even because each node is adjacent to k other nodes and has in-degree = out-degree = k. (An *Euler tour* is a path going through each arc exactly once and coming back to the same node.)
d. The diameter of $D(k, n)$ is n. (The diameter is the maximum distance between any pair of nodes.)
e. The connectivity of $D(k, n)$ is $k - 1$. (Connectivity measures the degree of fault tolerance and resilience.)
f. $D(k, n + 1)$ is the line digraph of $D(k, n)$. (Although $D(k, n)$ is not a subgraph of $D(k, n + 1)$, it is possible to identify some subgraph of $D(2, n)$ as a universal de Bruijn building block of order n for any de Bruijn digraph $D(2, N)$ with $n \leq N$.)

Because of these important properties, de Bruijn digraphs have been used as interconnection networks for computing, communications, and VLSI design. Hsu and Wei (1997) proposed efficient routing and sorting schemes for de Bruijn networks. Espona and Serra (1998) used de Bruijn digraphs to construct Cayley digraphs. Because de Bruijn digraphs are iterated line digraphs of complete symmetric digraphs (see Remark 2f), valuable information can be obtained about their Cayley regular covers regarding diameters, Hamiltonicity, fault tolerance, degrees of symmetry, and routings. For other parallel and distributed networks using de Bruijn and circulant digraphs, see Bermond et al. (1995) and Grammatikakis et al. (2001).

De Novo DNA Assembly Using Short Sequence Reads

Idury and Waterman (1995) were able to apply the Eulerian path approach to fragment assembly by combining the shotgun sequencing and sequencing by hybridization (SBH). Their approach entails the transformation of every read of length n into a collection of $[(n - k) + 1]$ k-tuples. By abandoning the OLC paradigm, they reduced the SBH and shotgun sequencing to an Eulerian path problem in a de Bruijn digraph. Although their approach was promising, it did not scale well because of many erroneous edges resulting from sequencing errors. In addition, it also contains plenty of repeats, which would make the de Bruijn digraph more entangled.

Pevzner et al. (2001) developed a practical fragment assembly software tool, called EULER, to address both shortcomings of the Idury–Waterman approach. The software produced error-free assemblies for several bacterial sequencing projects. Pevzner and Tang (2001) proposed a new EULER-DB algorithm that takes advantage of clone-end sequencing by using the double-barreled (DB) data. In addition, EULER-DB does not mask repeats but takes advantage of them as a powerful tool for **contig** ordering. Moreover, EULER-DB maps every read into some edges of the de Bruijn diagraph instead of nodes in traditional fragment assembly approaches.

Next-generation high-throughput sequencing became more popular in the 2000s with much shorter reads (25–400 bp vs. 500–750 bp), higher coverage (30× and higher vs. 10×), and lower cost. However, the existing assembly programs and error correction routine had not been modified to take advantage of the new technology for the assembly of short reads. Chaisson et al. (2004) presented a routine for base calling in reads before their assembly. They also showed that it is feasible to assemble such short reads that require a tremendous amount of finishing efforts to obtain the final contigs. They proposed an Eulerian assembler based on the concept of an A-Bruijn graph. The EULER+ assembler deals with errors in reads by inducing nodes with ungapped algorithms and by defining a set of graph-simplification methods that removes erroneous edges. Chaisson and Pevzner (2008) modified the A-Bruijn technique for short read assembly and presented a memory-efficient de Bruijn–based approach that supports A-Bruijn-like graph correction operations. The modified system, called EULER-SR, replaced the maximum spanning-tree optimization of the A-Bruijn graph by the maximum branching optimization on de Bruijn graphs.

We have seen so far that the de Bruijn digraph framework is ideal for handling high coverage and short sequencing reads and has many useful properties. For example, Eulerian paths in the de Bruijn digraph might give us a candidate for the original sequence. **Roche's 454 assembler**, called newbler, is the first assembler to exploit the short read technology and de Bruijn digraphs. In late 2007 and early 2008, several second-generation de Bruijn digraph assemblers were released for short reads compatible with the Illumina technology. These include EULER-SR (Chaisson and Pevzner 2008), Velvet (Zerbino and Birney 2008), ALLPATHS (Butler et al. 2008), and ABySS (Simpson et al. 2009), which are explicitly based on de Bruijn digraphs. They also include SSAKE (Warren et al. 2007), VCAKE (Jeck et al. 2007), and SHARCGS (Dohm et al. 2007), which are implicitly based on de Bruijn digraphs and use a pre-tree-based approach.

Focusing on the **Illumina sequencing** platform, Hernandez et al. (2008) presented an assembler called Edena (Exact De Novo Assembler), which is based on the classical Overlay-Layout-Consensus (OLC) assembly model. They applied the system to two data sets from the *Staphylococcus aureus* strain MW2 genome and the *Helicobacter acinonychis* strain Sheeba genome. Recognizing significant overlaps between contigs produced by Edena and Velvet, they obtained two additional data sets by combining the contigs generated by the two programs. They showed that the two systems are partially complementary and that their combined usage can lead to longer contigs.

Genome Assembly Using de Bruijn–Based and Generalized de Bruijn Digraphs

The genome assembly problem can be traced back to the 1980s on DNA fragment assembly using the OLC algorithm. It has evolved for the last three decades and was impacted by the shotgun sequencing and sequencing by hybridization (SBH) in the 1990s. Owing to the advent of high-throughput sequencing technology, de novo assembly using short sequence reads has become not only popular but also necessary. Computational efficiency has been improved by using Message-Passing Interface (MPI)–cluster approach (e.g., ABySS) (Simpson et al. 2009). The general assembly systems and these large computation techniques will certainly become more widespread and mature in the next years. Flicek and Birney (2009) provided an excellent review on the system and algorithm approaches and offered future directions for these tools.

We note that even though the many genome assembly algorithms (including EULER-DB, EULER-SR, ALLPATHS, Velvet, ABySS, SSAKE, SHARCGS, and VCAKE) implicitly or explicitly involve de Bruijn digraphs, they essentially use the concept of a de Bruijn digraph interconnection rather than the de Bruijn digraph

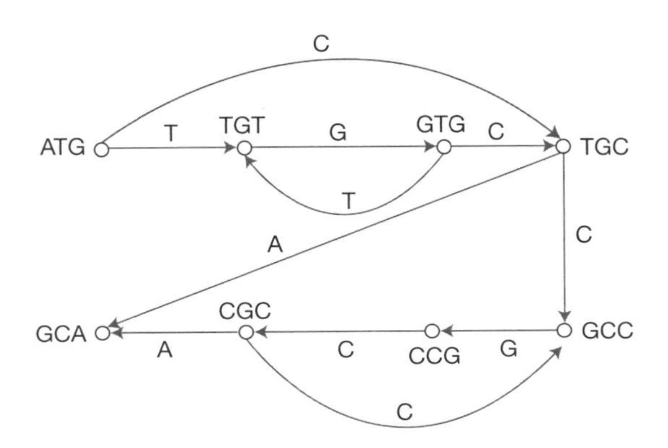

FIGURE 3. de Bruijn–based digraph $D(3; 8, 11)$ from the DNA fragments in F = {ATG, GCA, TGT, GCC, GTG, CCG, TGC, CGC}.

as defined in the mathematical community. As such, we feel the need to define the concept of a de Bruijn–based digraph $D(k; p, e)$ of k-tuples (or sequence of length k) with $k \leq n$, the length of DNA fragments.

Definition 1: Let k be a positive integer less than or equal to the minimum length of DNA fragments. Formally, for a fixed k, a "sequence graph" is a directed graph whose nodes are sequences, having the property that whenever $x \rightarrow y$ is an edge, the end of the first node perfectly overlaps the beginning of the second node by exactly $k - 1$ bases. The *de Bruijn–based (DBB) digraph* $D(k; p, e)$ is a directed graph with p nodes, where each node is a sequence of alphabets in {A, C, G, T} with length k, written as $b_1 b_2 \cdots b_k$ and e edges. Each edge is a connection from node $b_1 b_2 \cdots b_k$ to node $b_2 b_3 \cdots b_k \alpha$ for some α in {A, C, G, T}.

One example is the case in which the original genome (or sequence) is S_1 = ATGTGCCGCA and the set of DNA fragments from the SBH is F = {ATG, GCA, TGT, GCC, GTG, CCG, TGC, CGC}. A de Bruijn–based digraph $D(3; 8, 11)$ can be constructed (Fig. 3). The second example is the DBB digraph $D(2; 7, 8)$, which was the spectrum graph obtained from the seven substrings (each with length = 2) in F studied by Idury and Waterman (1995) (see Fig. 4). The only Eulerian path (or cycle) in $D(2; 7, 8)$ in Figure 4 would lead to the original sequence S_1.

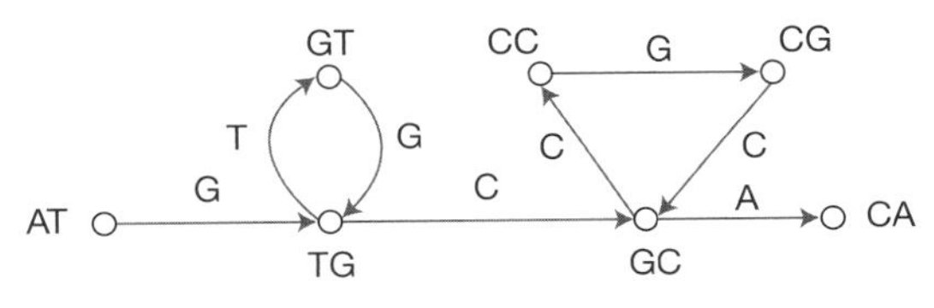

FIGURE 4. de Bruijn–based digraph $D(2; 7, 8)$ corresponding to the genome sequence S_1 = ATGTGCCGCA.

The third example is the construction and visualization of a de Bruijn digraph from a DNA sequence by Flicek and Birney (2009, Fig. 3 on p. S10). The original sequence

$$S_2 = \text{TAGTCGAGGCTTTAGATCCGATGAGGCTTTAGAGACAG}$$

has length $q = 38$ and a set of $m = 63$ fragments each with length $n = 7$. The set of 252 4-mers requires a de Bruijn–based diagraph $D(k; p, e)$ with $k = 4$, $p = 37$ and possible repeats at each node, and $e = 36$ arcs.

We note that the DBB digraph $D(2; 7, 8)$ in Figure 4 is simple and contains a single Eulerian path that would lead to the original sequence S_1. But the DBB digraph $D(4; 37, 36)$ (Flicek and Birney 2009) has several linear stretches, bubbles, and tips (a node with in-degree = 1 and out-degree = 0). After removing those bubbles and tips, the digraph will turn out to be in a different format, which we define here.

Definition 2: Let S be a DNA or genome sequence of length q with each component in {A, C, G, T}. *A generalized de Bruijn (GDB) digraph* $G(p', e'; K, T, H)$ is a directed graph with p' nodes labeled as $b_1b_2\cdots b_k$ for some k in $K = \{k_1, k_2, \ldots, k_n\}$ and $k_i < q$ and e' arcs labeled as $(t, c_1c_2\cdots c_k)$ for some t in $T = \{t_1, t_2, \ldots, t_r\}$, $t_i \leq k - 1$, and some h in $H = \{h_1, h_2, \ldots, h_s\}$, $h_i < q$, so that node $b_1b_2\cdots b_k$ is adjacent to $b_{t+1}b_{t+2}\cdots b_kc_1c_2\cdots c_h$, where b_i, c_i are elements in {A, C, G, T}.

We note that the GDB digraph thus defined has nodes labeled as sequences of variable length k (k in K), node $b_1b_2\cdots b_k$ can be shifted to the left t spaces (t in T) (vs. only $t = 1$ for the DBB digraph D defined in Definition 1), and a sequence $c_1c_2\cdots c_h$ of various length h (h in H) can be concatenated from the right-hand side. The DBB digraph $D(4; 37, 36)$ obtained above from the original unknown sequence S_2 will become, after removal of bubbles and tips and simplification of line stretches, a GDB digraph $G = G(4, 4; K, T, H)$, $K = \{8, 11, 12\}$, $T = \{5, 8, 9\}$, and $H = \{5, 8, 9\}$ as follows (Flicek and Birney 2009, Figs. 3a, 4) in Figure 5.

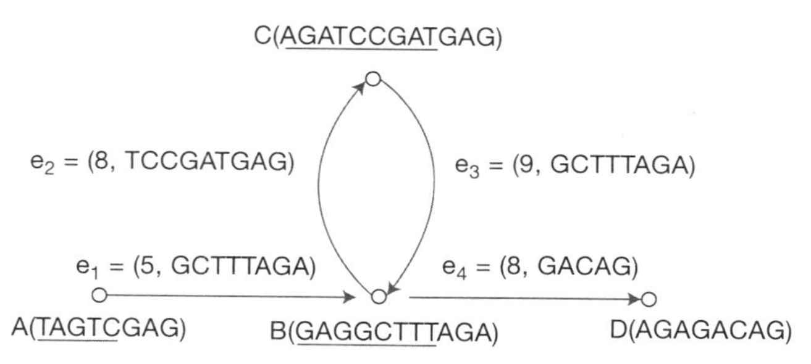

FIGURE 5. The GDB digraph $G(4, 4; K, T, H)$, where $K = \{8, 11, 12\}$, $T = \{5, 8, 9\}$, and $H = \{5, 8, 9\}$, derived from the DBB digraph $D(4; 37, 36)$ obtained by Flicek and Birney (2009).

From this GDB digraph $G(4, 4; K, T, H)$, the following Eulerian path $A\ e_1e_2e_3e_4$, which is the concatenation of string A and a sequence of strings labeled as consecutive edges in the Eulerian path, would lead to the original sequence S_2 of length 38:

$$A \xrightarrow{e_1} B \xrightarrow{e_2} C \xrightarrow{e_3} B \xrightarrow{e_4} D$$

With the de Bruijn–based (DBB) digraph $D(k; p, e)$ and the generalized de Bruijn (GDB) digraph $G(p', e'; K, T, H)$ well defined in Definitions 1 and 2, we are ready to describe a general genome assembly framework using a DBB digraph and a GDB digraph as follows:

Remark 3: General Genome Assembly Framework (GAF). Let S be a DNA sequence or unknown genome to be assembled, $S = a_1a_2...a_q$, where each a_i is in {A, C, G, T}. The following procedure illustrates a general genome assembly framework using de Bruijn–based (DBB) and generalized de Bruijn (GDB) digraphs.

1. Use various techniques to break genomic DNA to create multiple random overlapping fragments of genome S. Use NGS to determine the sequence of a set of m DNA reads: $F = \{f_1, f_2, \ldots, f_m\}$ with each read length $|f_i| = n$.
2. Apply a hash function or other mapping technique to F to define all k-tuples. Output a de Bruijn–based (DBB) digraph of k-tuples, $D(k; p, e)$ (see Definition 1), with p nodes and e arcs, where $k \leq n$.

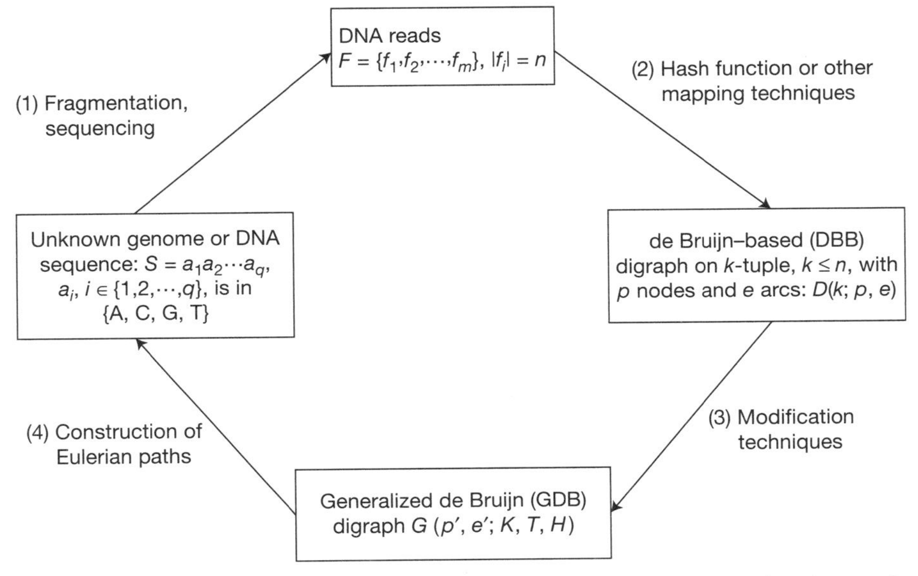

FIGURE 6. A general Genome Assembly Framework (GAF) using de Bruijn–based and generalized de Bruijn digraphs.

3. Apply any necessary modification techniques to output a generalized de Bruijn (GDB) digraph $G(p', e'; K, T, H)$ with p' nodes and e' arcs (see Definition 2).
4. Construct an Eulerian path in $G(p', e'; K, T, H)$. Output the path and compare them if there are more than one.

The steps of the general genome assembly framework in Remark 3 are also illustrated in Figure 6.

REFERENCES

Bermond J-C, Comellas F, Hsu DF. 1995. Distributed loop computer networks: A survey. *J Para Dist Comput* **24:** 2–10.

Butler J, MacCallum I, Kleber M, Shlyakhter IA, Belmonte MK, Lander ES, Nusbaum C, Jaffe DB. 2008. ALLPATHS: de novo assembly of whole-genome shotgun microreads. *Genome Res* **18:** 810–820.

Chaisson MJ, Pevzner PA. 2008. Short read fragment assembly of bacterial genomes. *Genome Res* **18:** 324–330.

Chaisson M, Pevzner PA, Tang H. 2004. Fragment assembly with short reads. *Bioinformatics* **20:** 2067–2074.

de Bruijn NG. 1946. *A combinatorial problem*, Vol. 49, pp. 758–764. Koninklijke Nederlandse Akademie v. Wetenschappen, Amsterdam.

Dohm JC, Lottaz C, Borodina T, Himmelbauer H. 2007. SHARCGS, a fast and highly accurate short-read assembly algorithm for de novo genomic sequencing. *Genome Res* **17:** 1697–1706.

Espona M, Serra O. 1998. Cayley digraphs based on the de Bruijn networks. *SIAM J Discrete Math* **11:** 305–317.

Fiol MA, Yebra JLA, Alegre I. 1984. Line digraph iterations and the (d, k) digraph problem. *IEEE Trans Comput* **C-33:** 400–403.

Flicek P, Birney E. 2009. Sense from sequence reads: Methods for alignment and assembly. *Nat Methods* **6:** S6–S12.

Grammatikakis MD, Hsu DF, Kraetzl M. 2001. *Parallel system interconnections and communications*. CRC Press, Boca Raton, FL.

Hernandez D, François P, Farinelli L, Osterås M, Schrenzel J. 2008. De novo bacterial genome sequencing: Millions of very short reads assembled on a desktop computer. *Genome Res* **18:** 802–809.

Hsu DF, Wei DSL. 1997. Efficient routing and sorting schemes for de Bruijn networks. *IEEE Trans Parallel Distributed Sys* **8:** 1157.

Idury RM, Waterman MS. 1995. A new algorithm for DNA sequence assembly. *J Comput Biol* **2:** 291–306.

Imase M, Itoh M. 1981. Design to minimize diameter on building-block network. *IEEE Trans Comput* **C-30:** 439–442.

Jeck WR, Reinhardt JA, Baltrus DA, Hickenbotham MT, Magrini V, Mardis ER, Dangl JL, Jones CD. 2007. Extending assembly of short DNA sequences to handle error. *Bioinformatics* **23:** 2942–2944.

Lempel A. 1971. m-ary closed sequences, *J Comb Theory* **10:** 253–258.

Pevzner PA, Tang H. 2001. Fragment assembly with double-barreled data. *Bioinformatics* **17:** S225–S233.

Pevzner PA, Borodovsky MY, Mironov AA. 1989. Linguistics of nucleotide sequences. II: Stationary words in genetic texts and the zonal structure of DNA. *J Biomol Struct Dyn* **6:** 1027–1038.

Pevzner PA, Tang H, Waterman MS. 2001. An Eulerian path approach to DNA fragment assembly. *Proc Natl Acad Sci* **98:** 9748–9753.

Reddy SM, Pradhan DK, Kuhl JG. 1980. *Directed graphs with minimal diameter and maximum node connectivity*. School of Engineering Oakland University Technical Report.

Simpson JT, Wong K, Jackman SD, Schein JE, Jones SJ, Birol I. 2009. ABySS: A parallel assembler for short read sequence data. *Genome Res* **19:** 1117–1123.

Staden R. 1980. A new computer method for the storage and manipulation of DNA gel reading data. *Nucleic Acids Res* **8:** 3673–3694.

Warren RL, Sutton GG, Jones SJ, Holt RA. 2007. Assembling millions of short DNA sequences using SSAKE. *Bioinformatics* **23:** 500–501.

West DB. 2001. *Introduction to graph theory*, 2nd ed. Prentice-Hall, Upper Saddle River, NJ.

Zerbino DR, Birney E. 2008. Velvet: Algorithms for de novo short read assembly using de Bruijn graphs. *Genome Res* **18:** 821–829.

6

De Novo Assembly of Bacterial Genomes from Short Sequence Reads

Silvia Argimón and Stuart M. Brown

Whole-genome sequencing is at the core of comparative genomic studies that aim to identify the genetic basis of bacterial virulence. **Next-generation sequencing** (NGS) platforms now make the comparison of a large number of bacterial genomes affordable even outside of the large genome research centers. The new field of genomic epidemiology makes use of whole-genome sequencing of multiple isolates to reconstruct disease outbreaks, and it is becoming an important part of public health surveillance and disease response efforts.

In 2002 The Institute for Genomic Research (TIGR) spent several months comparing the whole genome of anthrax spores from letters mailed to a member of the U.S. Congress to the genome sequences of known strains. Sequencing of a total of 16 anthrax strains was conducted by the **Sanger sequencing method** on ABI 3730xl96-capillary machines. From 60,000 to 100,000 **sequence reads** were collected from each strain, providing a minimum of 12× **coverage**. This represents approximately 10,000 individual runs of the ABI 3730xl machines (Rasko et al. 2011). Specific sequence markers were identified in isolates found in the letters, which played an important role in the forensic investigation.

With the availability of NGS, whole-genome sequencing of multiple strains of a pathogen is becoming common practice. A study of an outbreak of methicillin-resistant *Staphylococcus aureus* (MRSA) in a London hospital included whole-genome sequences of 63 isolates (Harris et al. 2010). Similarly, Canadian researchers sequenced the whole genomes of 36 isolates of *Mycobacterium tuberculosis* to characterize an outbreak in British Columbia in 2007 (Gardy et al. 2011). Investigators from the U.S. Food and Drug Administration used whole-genome NGS of 35 isolates of *Salmonella* to track an outbreak of food poisoning in peppers used to season spiced meats (Lienau et al. 2011). The FDA will continue to use routine NGS to track *Salmonella* and other foodborne pathogens. Investigators at the Methodist

Hospital Research Institute of Houston sequenced the genomes of a total of 301 isolates of infectious Group A *Streptococcus* (Shea et al. 2011). For each of these studies, sequencing was conducted to a depth of 50× –100× coverage for each strain.

As routine whole-genome sequencing of bacterial strains becomes affordable, researchers are faced with significant bioinformatics challenges. These massively parallel sequencing platforms, such as the Illumina Genome Analyzer, typically produce high-throughput, high-quality **short reads**, which has prompted the development of **de novo sequence** assemblers specifically suited for this type of sequence data (see Chapter 5). However, the literature benchmarking and comparing the use of the different assembly **algorithms** is scant, and the inexperienced user faces the assembly task aided only by the software's *User Manual*, the information in online forums, and a few publications for the mathematically gifted.

In this section, we present the assembly of *Streptococcus mutans* genomes (genome size 2.03 Mb) from 20 clinical isolates, focusing on a practical comparison of two **de novo assemblers** for short sequence data: Velvet (Zerbino and Birney 2008) and ABySS (Simpson et al. 2009). Both of them are based on a **de Bruijn graph** approach, where elements are organized around words of *k* nucleotides, or ***k*-mers**. In addition, we present two methods to evaluate de novo genome assemblies: mapping **contigs** to the known genome sequence of a related organism using Mauve Contig Mover, and aligning contigs to the sequence of a known gene.

For our comparative genomics study of *S. mutans*, 20 samples were multiplexed, and library preparation was done using the Illumina TruSeq DNA Sample Prep Kit. Libraries were run on the Illumina HiSeq 2000 to yield 50-bp **paired-end reads**. The sequence reads of each sample were contained in two **FASTQ files**.

VELVET

The Velvet software package is accompanied by a manual that contains all of the basic running information in an easy-to-follow fashion. A Velvet assembly is achieved in two consecutive commands: `velveth` and `velvetg`. Velveth (Velvet hash) reads the sequence input files and outputs three files, Sequences, Roadmaps, and Log. Velvetg (Velvet graph) uses these files to build the assembly and outputs the files contigs.fa, UnusedReads.fa (if specified), Graph2, LastGraph, PreGraph, and stats.txt and will also write to the Log file. Velveth requires an output directory, the *k*-mer length (must be an odd number), the sequence file format, the read type, and the input filename(s). For example, for our sample 04 and a $k = 31$:

```
>velveth velvet_output 31 – fastq
– shortPaired s_8_1_sequences.txt s_8_2_sequences.txt
```

Velvetg requires the coverage cutoff to be specified in order to exclude short, low-coverage nodes from the assembly. In addition, running `velvetg` on paired-end reads requires the expected insert length (the average length of the sequenced fragment) and the expected *k*-mer coverage. For example, for our samples, we expected coverage around 100× and we estimated our insert length to be 164 nucleotides based on the quality data supplied by the sequencing facility. We set the coverage cutoff value to auto, so that `velvetg` estimates it:

```
>velvetg velvet_output – ins_length 164 – exp_cov 100
– cov_cutoff auto
```

In addition, the user can request that the contigs in the contigs.fa file be longer than a given value, and to create a file with all the reads that are not included in the assembly. We requested that the contigs in the contigs.fa file be longer than 100 nucleotides. Thus, for sample 04:

```
>velvetg velvet_output – min_contig_lgth 100 – ins_length
164 – exp_cov 100 – cov_cutoff auto – unused_reads yes
```

The Log file summarized the output as

```
Final graph has 426 nodes and n50 of 28569, max 66443,
total 1939136, using 9358159/9483822 reads
```

This means that the assembly contains 426 contigs, the median length-weighted contig length is 28,569 nucleotides, the biggest contig is 66,443 nucleotides, the coverage of the assembly totals 1.94 Mb, and the assembly was built with 98.67% of the reads.

Effect of the Insert Length on the Assembly

In addition to the estimated insert length of 164 nucleotides, we also tested an insert length of 283, and not specifying any insert length, which prompts `velvetg` to measure it automatically based on the read-pairs information. In this case `velvetg` estimated the insert length to be 228 nucleotides, which is 64 nucleotides larger than our estimation based on the library fragment size data supplied by the sequencing facility.

Table 1 shows that Velvet yielded a better assembly when no insert length was specified (i.e., when `velvetg` estimated the insert length to be 228 nucleotides).

TABLE 1. Effect of the insert length on the Velvet assembly of sample 04 reads

	-ins_length 164	-ins_length 283	No insert length specified (228)
Number of nodes	426	417	183
N_{50}	28,569	28,569	1,102,815
Max	66,443	66,443	1,102,815
Total	1,939,136	1,939,177	1,968,956
Used reads	98.67%	98.68%	98.73%

The assembly parameters were the same, with the sole exception of the insert length: velveth was computed for k = 25; velvetg parameters were min_contig_lgth 100 and exp_cov auto.

Effect of the Expected Coverage on the Assembly

We tested four different expected k-mer coverage values: 50, 100, 200, and the "auto" option, by which `velvetg` sets the expected coverage to the length-weighted median contig coverage and the coverage cutoff to half that value. In this case, `velvetg` estimated the coverage to be 115.76, which was not too far from our coverage estimate of 100.

Table 2 shows that the expected coverage value greatly influences the final assembly. Our estimate of the coverage was rather accurate, and therefore the final assembly values were similar to those generated by the "auto" option.

Effect of the *k*-mer Size on the Final Assembly

By default, the maximum k-mer length is 31, but this limitation can be overcome during compilation, which is convenient when assembling reads longer than 36 bp. Multiple k-mer lengths can be tested in one `velveth` command to avoid

TABLE 2. Effect of the expected coverage value on the Velvet assembly of sample 04 reads

	-exp_cov 50	-exp_cov 100	-exp_cov 200	-exp_cov auto
Number of nodes	434	195	167	183
N_{50}	28,569	1,796,683	401,886	1,102,815
Max	66,443	17,996,683	790,217	1,102,815
Total	1,939,052	1,979,491	1,949,637	1,968,956
Used reads	98.64%	98.75%	98.73%	98.73%

The assembly parameters were the same, with the sole exception of the expected coverage: velveth was computed for k = 25; velvetg parameters were min_contig_lgth 100, cov_cutoff auto, and no insert_length was specified (auto = 228).

TABLE 3. Effect of the *k*-mer size on the Velvet assembly of sample 04 reads

k	Number of nodes	N_{50}	N_{max}	Total	% used reads
25	183	1,102,815	1,102,815	1,968,956	98.73
27	143	1,722,193	1,722,193	1,972,280	98.20
29	128	1,072,586	1,072,586	1,977,225	97.84
31	110	1,948,103	1,948,103	1,965,894	96.89
33	455	15,367	82,041	1,961,880	96.05
35	135	1,112,831	1,112,831	2,141,512	96.08
37	165	1,009,775	1,240,704	2,481,693	95.29
39	292	729,917	833,798	2,690,036	92.64
41	474	12,952	55,310	2,606,831	92.07
43	457	12,451	76,281	2,659,230	91.41
45	56	1,044,727	1,044,727	1,968,694	93.04

repetitive computations. For example, for $25 \geq k \leq 45$, in increments of 2,

```
>velveth velvet_output 25, 47, 2 – fastq
– shortPaired s_8_1_sequences.txt s_8_2_sequences.txt
```

Note that specifying *k* as `25,45,2` will run `velveth` for $25 \geq k \leq 43$, not $25 \geq k \leq 45$. This creates a subdirectory velvet_output_k$_i$ for each *k* value tested. Running `velvetg` for each one of these directories, we get

```
>velvetg velvet_output_ki – min_contig_lgth 100 – exp_cov auto
– unused_reads yes
```

Note that when the expected coverage is set to "auto," the coverage cutoff is also set to "auto" without the need to specify it. This means that `velvetg` will set the `cov_cutoff` to half the length-weighted median contig coverage depth, which is not necessarily the same for all *k*-mer sizes. In our experience, performing a *k*-mer size optimization while setting other velvetg parameters to "auto" yields better assemblies.

Table 3 shows that the choice of *k*-mer size has a large effect on the assembly. The number of nodes, the N_{50}, or the total coverage do not seem to follow any mathematical function (Fig. 1). Unexpectedly, the coverage (total) for $k = 37–43$ far exceeds the average size of the known *S. mutans* genomes (~2.0 Mb), which suggests an artifact in genome assembly. Zerbino and Birney (2008) recommend choosing the assembly that produces the highest N_{50} contig length. According to this guideline, the assembly obtained for $k = 31$ would be the optimal one.

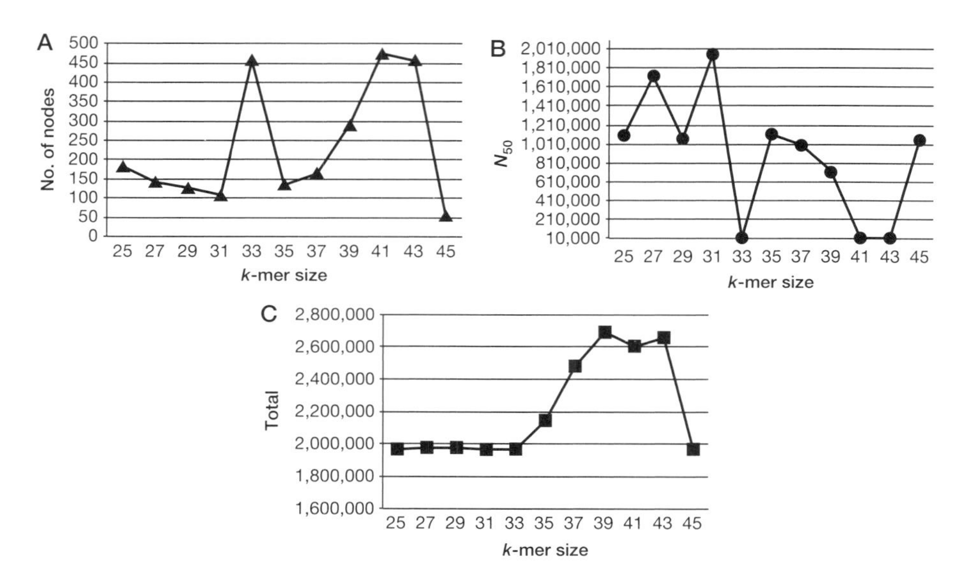

FIGURE 1. Variation of Velvet assembly values with *k*-mer size. (*A*) The number of nodes (contigs) produced by assembly with Velvet at different *k*-mer size values do not follow any consistent trend or predictable pattern. (*B*) The N_{50} value of contigs produced by assembly with Velvet also show no clear pattern. (*C*) The total length of all contigs (sum) increases to a maximum for *k*-mers of 39–43 but declines at 45.

VelvetOptimiser

The Velvet software package includes the script VelvetOptimiser, which uses a heuristic method to find the optimal *k*-mer length and coverage cutoff for Velvet. We tested this script for assembly of our sample 04 sequence reads as follows:

```
> VelvetOptimiser.pl − s 25 − e 45 − f′ − shortPaired
− fastq s_8_1_sequence.txt s_8_2_sequence.txt′
```

The options -s and -e indicate the lower and the higher *k*-mer size to be tested, respectively. The option -f is the file section of the `velveth` command line. In this case, VelvetOptimiser set the insert length to auto, and found the optimal cutoff value to be 0.91 and the optimal *k*-mer length to be 43. The assembly files were created in a subdirectory called auto_data. Note that if you run several instances of VelvetOptimiser, you will need to rename the auto_data subdirectory or it will be overwritten.

By default, VelvetOptimiser will choose the optimal *k*-mer size based on the N_{50}. However, the – k option enables the user to base the assembly optimization function on other variables such as the total number of base pairs in contigs (see Table 4):

```
> VelvetOptimiser.pl − s 25 − e 45 − k tbp − f′ − shortPaired
− fastq s_8_1_sequence.txt s_8_2_sequence.txt′
```

TABLE 4. Comparison of VelvetOptimiser results obtained with two different criteria of optimal *k*-mer choice

	-k n50 (default)	-k tbp
Optimal *k*	43	25
Number of nodes	149	93
N_{50}	54,800	1,965,578
N_{max}	162,596	1,965,578
Total base pairs in contigs	1,966,327	1,986,015
Number of contigs > 1 kb	63	5
Total base pairs in contigs > 1 kb	1,948,636	1,975,426
Optimal cutoff	0.91	19.53
Expected coverage	33	112

Running VelvetOptimiser with the total number of base pairs in contigs as the optimization function yielded an optimal assembly for $k = 25$, containing a very large contig of ~1.97 Mb (i.e., almost the size of the entire genome). Interestingly, the N_{50} is much larger than the one obtained for the default settings, which according to the VelvetOptimiser manual uses the N_{50} as the optimization function. Moreover, the number of nodes, the N_{50}, and the N_{max} obtained with the VelvetOptimiser are very different from those obtained for $k = 43$ and $k = 25$ in Table 3, even though the options were set to auto.

In our experience, obtaining an optimal assembly with Velvet requires the optimization of the different input parameters, and it is advisable to set these to auto to compare with the user's estimations. VelvetOptimiser also appears to be a useful way to find the optimal assembly.

ABySS

ABySS (Assembly By Short Sequencing) allows for the computation of the assembly in a parallel environment, which can be an advantage over Velvet when working with large genomes. Because the ABySS software package is not accompanied by a detailed manual, the new user must rely on the README file and MAN pages and a support forum for running instructions. An ABySS assembly can be obtained with only one command that requires the *k*-mer length (*k*), the input sequence file(s), and the name of the output file. For paired-end reads, the minimum number of pairs required to join two contigs (*n*) also needs to be specified. For example, for our sample 04:

```
> abyss-pe k = 31 n = 10 in = 's_8_1_sequence.txt
s_8_2_sequence.txt' name = samp04
```

This command will output four files: samp04-contigs.fa contains the contigs, samp04-bubbles.fa contains variant sequences equal length, samp04-indel.fa contains variant sequences of different lengths, and samo04-contigs.dot indicates which contigs overlap and by how much. The rest of the files represent intermediate stages of ABySS.

ABySS will not automatically output a file with assembly statistics like Velvet's Log file. A stats file can be generated with the `abyss-fac` command, which is not described in the README file or MAN page but can be found in the ABySS forum:

```
>abyss-fac samp04-contigs.fa > samp04-stats.txt
```

An output filename needs to be provided or the output will only be directed to the screen:

n	n:100	n:N50	min	median	mean	N50	max	sum	
158	79	8	103	6116	25714	73111	227217	2031449	samp04-contigs.fa

`abyss-fac` computes these statistics for the contigs larger than 100 nucleotides. We achieved a similar situation with Velvet with the option `-min_contig_lgth 100`.

Thus, we computed the assembly with Velvet and ABySS using roughly the same parameters. We chose a *k*-mer size of 31 nucleotides ($k = 31$) and a minimum of 10 pairs to join two contigs together ($n = 10$ in ABySS, and default option for Velvet; see the *User's Manual*, Section 3.5, Advance Parameters: Pebble). ABySS does not require the expected coverage, the insert length, and the coverage cutoff, thus computing the Velvet assembly with the auto option for these parameters should yield similar results. A comparison of the Velvet assembly Log in Table 3 with the ABySS stats file shown in the box above suggests that the Velvet assembly was characterized by a higher N_{50} and N_{max}, but the ABySS assembly yielded a higher coverage, represented by the "sum" value. ABySS calculates these stats for contigs larger than 100 nucleotides (unless otherwise specified with the option `-t` of `abyss-fac`), thus the Velvet option `-min_contig_lgth 100` should be specified to be able to compare the values calculated by Velvet (Log) and ABySS (stats.txt).

Effect of *k* and *n* on the Final Assembly

Both the *k*-mer size (*k*) and the minimum number of pairs required to join two contigs (*n*) need to be optimized empirically. The default maximum value for *k* is 64, but this limit can be modified during compilation. We tested in parallel *k* values from 25 to 45 and five different values of *n* from 4 to 40. Similarly to what we observed

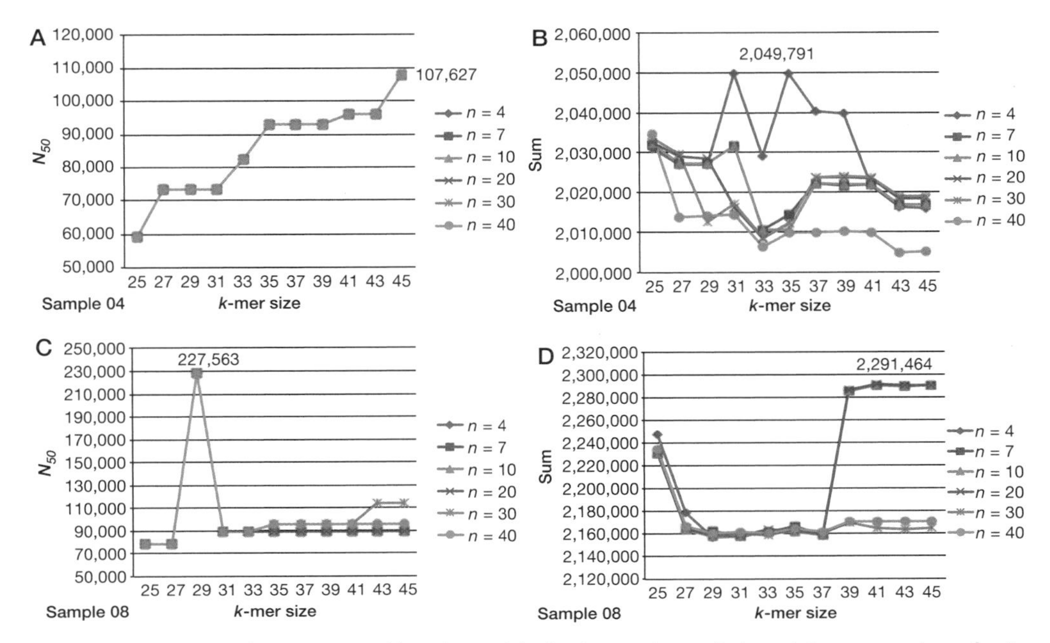

FIGURE 2. Variation of ABySS assembly values with the *k*-mer size and the minimum number of pairs required to join two contigs (from $n = 7$ to $n = 40$). The assembly N_{50} (*A*) and coverage (*B*) are shown for sample 04 and for sample 08 (*C,D*).

for Velvet, the N_{50} and the total coverage may behave rather erratically in response to different k values, hence the need to optimize empirically.

Our results show that the k value has a stronger effect on the final assembly than the n value, because the N_{50} was virtually the same for all n values tested (Fig. 2A,B). Moreover, the optimal k may vary greatly from sample to sample, and therefore it is advisable to optimize this value for each sample.

ASSEMBLY QUALITY

The optimal ABySS assembly for sample 04 based on the N_{50} value was found for $k = 45$. The N_{50} was exactly the same for all n values tested ($k = 45$), but the final assembly coverage was slightly higher for $n = 30$. On the other hand, the optimal Velvet assembly was obtained for $k = 31$, although we subsequently generated assemblies of even higher N_{50} values with VelvetOptimiser. The Velvet assemblies were all characterized by a higher N_{50} but a lower coverage than the ABySS assembly (Tables 4 and 5).

The assembly values such as N_{50}, number of contigs, and coverage are informative when trying to choose the best assembly possible, but they provide limited information regarding the assembly correctness. Moreover, the contigs represent puzzle

TABLE 5. Comparison of optimal assemblies obtained with Velvet and ABySS for sample 04

	Velvet	ABySS
k	31	45
Number of contigs	110	82
N_{50}	1,948,103	107,627
N_{max}	1,948,103	288,371
Coverage	1,965,894	2,018,686

pieces that need to be ordered into larger supercontigs or scaffolds to facilitate comparisons between genomes. If the genome sequence of a related organism is available, this information can be used to link the contigs together.

We built scaffolds from the Velvet and ABySS contigs with the Mauve Contig Mover (MCM) (Rissman et al. 2009), which ordered the contigs based on comparison with the **reference genome** of *S. mutans* strain UA159. Matches between the contigs and the reference genome are sorted into locally collinear blocks (LCBs). Each LCB represents a region of homologous sequence without rearrangement between the contigs and the reference genome. One contig can contain multiple LCBs, or one LCB may be split across two or more contigs. However, false rearrangements can result from misassemblies.

MCM was originally designed to make comparisons among multiple complete genomes to identify syntenic blocks. However, it works well as a scaffold builder and visualization tool for NGS assemblies of bacterial genomes. The Mauve program can be downloaded from the Genome Center of Wisconsin (http://gel.ahabs.wisc.edu/mauve/download.php) for a desktop computer (Mac, Windows, or Linux executables available). Launch the Mauve program and choose "Move Contigs" from the Tools menu. Load a FASTA file of a reference genome and a file with the set of contigs produced by Velvet or ABySS. (Note that an ABySS contig file requires some reformatting.) The first step in the MCM orders contigs by comparison with the reference genome and creates contig groups as LCBs—which are effectively the same as scaffolds. Subsequent **alignment** iterations shuffle and merge LCBs to match the reference genome until no further ordering is possible (see Fig. 3).

The panoramic views of the draft genomes represented by the Mauve alignments are a useful tool to assess the quality of the genome assembly. A different approach is to look at the coverage obtained for specific loci. The tandemly arranged *gtfB* and *gtfC* genes code for two glucosyltransferases that show great conservation, likely due to gene duplication. In addition, both enzymes contain a glucan-binding domain (GBD) composed of several direct repeats. Both of these characteristics could cause misassemblies around these loci. We aligned the Velvet and ABySS contigs to the *gtfB–gtfC* sequence from *S. mutans* strain UA159 with Sequencher (Gene

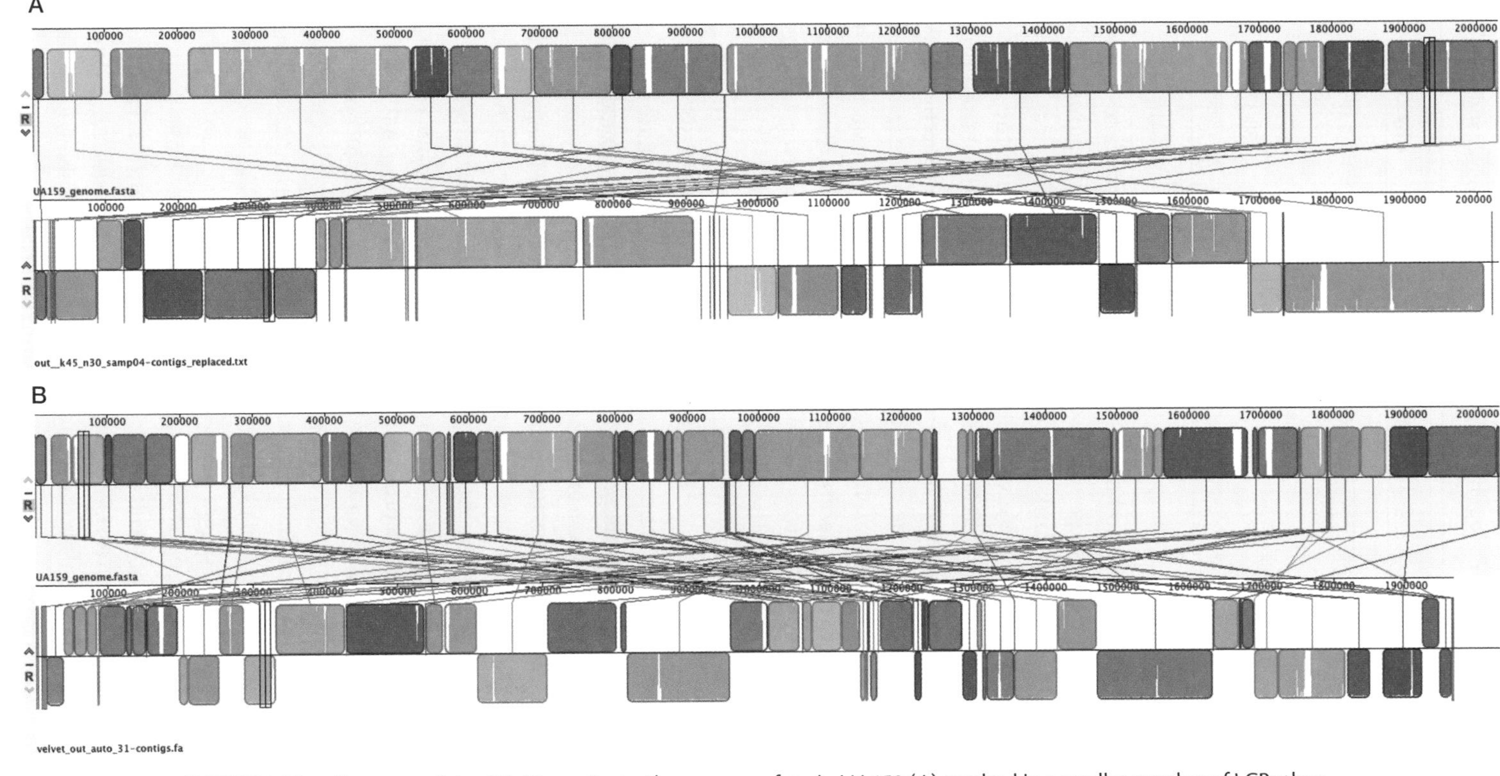

FIGURE 3. The alignment of the ABySS contigs to the genome of strain UA159 (*A*) resulted in a smaller number of LCBs than the alignment of the Velvet contigs to the same reference genome (*B*). This was the case for all of the Velvet assemblies, including those obtained with VelvetOptimiser (not shown). Assuming that the organization of the genomes between different strains of *S. mutans* is conserved, ABySS generated a more parsimonious assembly than Velvet did.

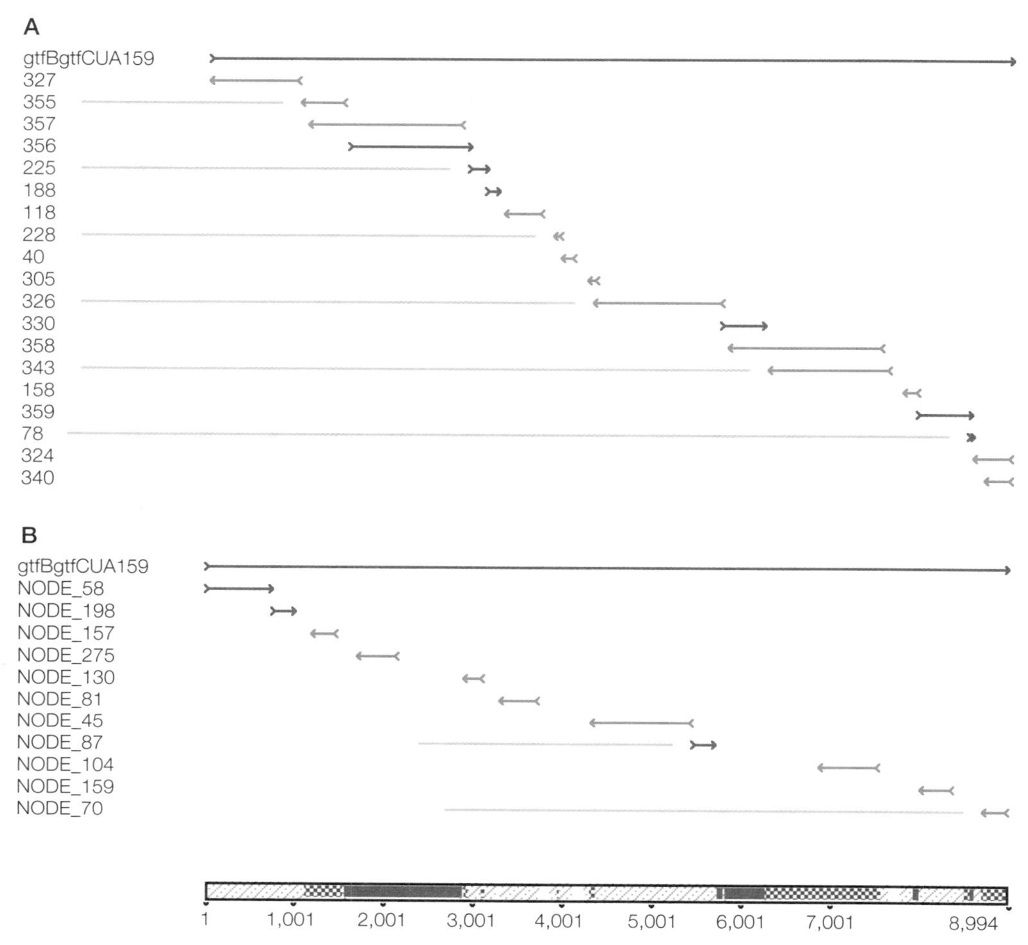

FIGURE 4. Coverage of the *S. mutans* UA159 *gtfB–gtfC* loci (represented by the full-length green arrow) by ABySS (*A*) or Velvet (*B*) contigs, assembled with Sequencher.

Codes) and found that this region of the genome was much better covered by the ABySS contigs than by the Velvet contigs. This suggests that the ABySS algorithm was more successful in assembling short reads into contigs that span a genome region potentially problematic for assembly (see Fig. 4).

CONCLUDING REMARKS

Our experience assembling bacterial genomes from short Illumina reads with two assembly tools, Velvet and ABySS, suggests that an assembly can be obtained with only a few commands. Obtaining a good quality assembly requires a significant effort in optimizing parameters and comparing the results. Moreover, the parameters need to be optimized for every sample, because the optimal values may vary unpredictably.

In our hands, Velvet seemed more user-friendly than ABySS, largely because the Velvet documentation is readily available for the first-time user. ABySS, however, generated a better assembly, as suggested by alignment of the contigs to an *S. mutans* reference genome, as well as by the coverage of the *gtfB–gtfC* loci.

REFERENCES

Gardy JL, Johnston JC, Ho Sui SJ, Cook VJ, Shah L, Brodkin E, Rempel S, Moore R, Zhao Y, Holt R, et al. 2011. Whole-genome sequencing and social-network analysis of a tuberculosis outbreak. *N Engl J Med* **364:** 730–739.

Harris SR, Feil EJ, Holden MT, Quail MA, Nickerson EK, Chantratita N, Gardete S, Tavares A, Day N, Lindsay JA, et al. 2010. Evolution of MRSA during hospital transmission and intercontinental spread. *Science* **327:** 469–474.

Lienau EK, Strain E, Wang C, Zheng J, Ottesen AR, Keys CE, Hammack TS, Musser SM, Brown EW, Allard MW, et al. 2011. Identification of a salmonellosis outbreak by means of molecular sequencing. *N Engl J Med* **364:** 981–982.

Rasko DA, Worsham PL, Abshire TG, Stanley ST, Bannan JD, Wilson MR, Langham RJ, Decker RS, Jiang L, Read TD, et al. 2011. *Bacillus anthracis* comparative genome analysis in support of the Amerithrax investigation. *Proc Natl Acad Sci* **108:** 5027–5032.

Rissman AI, Mau B, Biehl BS, Darling AE, Glasner JD, Perna NT. 2009. Reordering contigs of draft genomes using the Mauve aligner. *Bioinformatics* **25:** 2071–2073.

Shea PR, Beres SB, Flores AR, Ewbank AL, Gonzalez-Lugo JH, Martagon-Rosado AJ, Martinez-Gutierrez JC, Rehman HA, Serrano-Gonzalez M, Fittipaldi N, et al. 2011. Distinct signatures of diversifying selection revealed by genome analysis of respiratory tract and invasive bacterial populations. *Proc Natl Acad Sci* **108:** 5039–5044.

Simpson JT, Wong K, Jackman SD, Schein JE, Jones SJ, Birol I. 2009. ABySS: A parallel assembler for short read sequence data. *Genome Res* **19:** 1117–1123.

Zerbino DR, Birney E. 2008. Velvet: Algorithms for de novo short read assembly using de Bruijn graphs. *Genome Res* **18:** 821–829.

WWW RESOURCE

http://gel.ahabs.wisc.edu/mauve/download.php Mauve program download, Genome Evolution Laboratory, University of Wisconsin, Madison.

7

Genome Annotation

Steven Shen

BRIEF INTRODUCTION

What Is Genome Annotation?

Next-generation DNA sequencing (NGS) technology allows for the complete sequencing of novel genomes (**de novo sequencing**) much more rapidly and at a much lower cost than earlier **Sanger sequencing** techniques. However, the accelerated collection of sequence data creates a bottleneck for **assembly** of **short NGS reads** into entire chromosome sequences and the annotation of the new assemblies with gene features and other biologically relevant data. Some of these processes have been improved with new software, but others still require extensive manual efforts by skilled curators.

With respect to Stein's concept of genome annotation (Stein 2001), from a technical point of view, genome annotation is actually a process to structure the genome assembly, identify the gene models and genomic features, and make the connection between genome elements and biological meanings. This process is continuously evolving with new concepts of gene structure and function that emerge from advances in biological and medical research. The goal of genome annotation is to create an accurate and up-to-date coding reference for biological and medical research. This annotation is an essential prerequisite for other genomics technologies such as gene expression, alternative splicing, epigenetics, and the discovery of functional disease-causing mutations.

To better dissect the genome annotation process, researchers break down the analysis of genome sequences into three major internally connected parts: genome **sequence assembly**, assignment of genomic features (genes, non-coding RNAs, promoter regions, repeats, pseudogenes, etc.) to specific positions in the assembly, and assignment of biological traits (functions of each element). When a brand-new genome sequencing project starts, researchers will first take the genomic DNA from the species, sequence the whole genome, and **assemble** the **sequence fragments** to create **contigs** and scaffolds of this genome. When a draft assembly is available, as a second step, the curators begin working with the sequence to identify the genomic regions and features. In a narrow sense, this is the genome annotation process, which we are going to discuss further in this chapter. The third step is to connect biological information to genomic features, which can be performed simultaneously depending on the tools used for the second step.

Why Does Annotation Matter?

Genome assembly aligns the short sequence fragments in **FASTA format** into either chromosomes or scaffolds and contigs (see Chapter 6). A draft assembly can be imagined as a plain map without labels and legends, on which people won't be able to find places and directions. The annotation is to identify and position each genome element on the draft assembly as labels on the map. This will not just provide names, positions (start and end), and directions of genome features, but also allow researchers to do further work on it such as link the detailed biological information to the elements.

As stated above, genome annotation is a two-layer process—at the first layer, researchers manually or using computer programs mark and position the genome features, and at the following layer, the biological information is attached by either manual or computational methods (see Fig. 1).

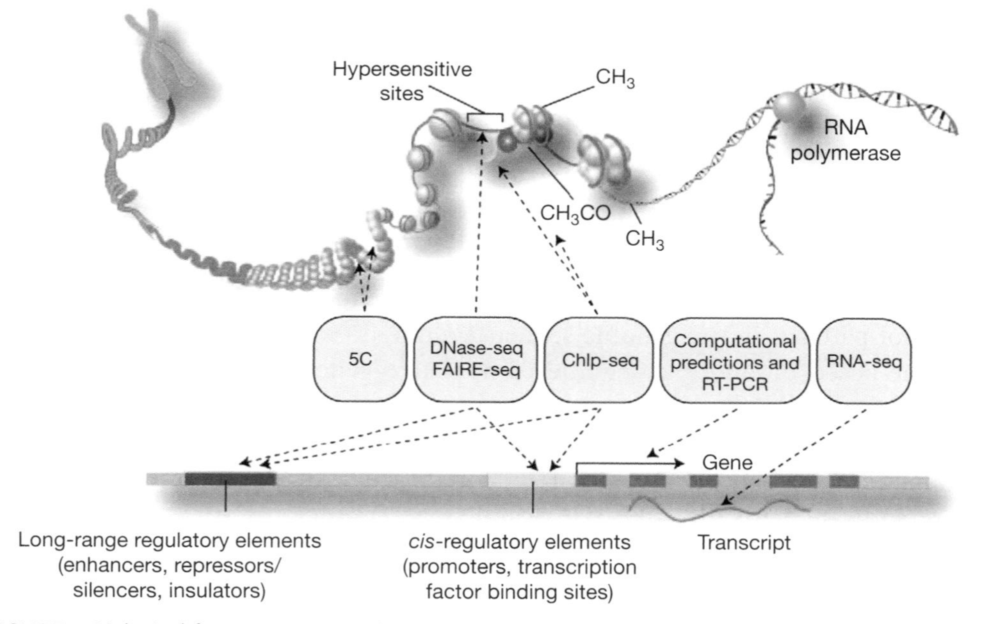

FIGURE 1. Biological features, genomic data sets (gray ovals), and computational predictions (colored bars). (Reprinted from The ENCODE Project Consortium 2011.)

Where Does Biological Annotation Information Come From?

A preliminary annotated assembly is a table format GTF or GFF file containing genome features with start and end positions with respect to a genomic **reference sequence** (a chromosome, contig, or scaffold) and some other basic genome specificities to which researchers can attach biological information. A partial GFF file that contains gene annotation information for one of the ant species is demonstrated in Table 1. At the present, there are some ongoing projects such as Ensembl, UniProt, RefSeq, and Gene Ontology that collect and store the genome information. These are publicly available databases that store the most up-to-date genome information and biological relevance for the most popular genomes.

There are many ways to retrieve the publicly available biological information from existing databases, which are detailed in the next section (and see Table 2). The simplest way is to do ortholog mapping, although this may only provide a fraction of gene model or genome feature information. HMMER, Blast2GO, and other programs are often used for function-based annotation. The function-based annotation requires fairly accurate gene model or structured genomic features. In the following sections, I mainly focus on how to build accurate gene models and genomic feature structures. The part of community-based annotation projects is listed in Table 3.

ANNOTATION STRATEGIES AND METHODS

There are two basic approaches to annotate a newly sequenced genome, the ab initio approach and the reference-based approach. The software programs and annotation strategies mentioned below are eukaryotic specific, unless otherwise noted.

Ab Initio Approach

Ab initio genome annotation methods identify genome features based on the recognition of patterns in the genome sequence alone no matter whether it is a single-genome-based or multiple-genome comparative genomics–based approach. The diversification of eukaryotic genome features, such as coding genes, **exons** and **introns**, non-coding RNAs, promoter regions, UTRs, repeats, pseudogenes, and so on, makes purely sequence-based prediction difficult. Although some other HMM-based software exists, the GENSCAN and GeneID programs are the most popular ones. With single-genome ab initio methods, only a fraction of true gene models (20%–30%) may be called from a new assembly in terms of prediction accuracy. The comparative genomics approach requires multiple genome assemblies of related species. This approach is able to improve annotation using conservation signatures

TABLE 1. An example of a GFF file

Chromosome ID	Annotation source	Feature type	Start	End	Score	Direction	Validation	X-ref
scaffold72	GLEAN	mRNA	211280	218219	0.973093	—	—	ID = Cflo_01967- -NA;
scaffold72	GLEAN	CDS	218078	218219	—	—	0	Parent = Cflo_01967- -NA;
scaffold72	GLEAN	CDS	212380	212681	—	—	2	Parent = Cflo_01967- -NA;
scaffold72	CUFF.196830.0	3′ UTR	211280	212379	—	—	—	Parent = Cflo_01967- -NA;
scaffold3033	GeneWise	mRNA	1399	2583	68.97	—	—	ID = Cflo_14312- -XP_001599370.1_NASVI;Shift = 0;
scaffold3033	GeneWise	CDS	1399	2583	—	—	0	Parent = Cflo_14312- -XP_001599370.1_NASVI;
scaffold859	GLEAN	mRNA	444197	446169	0.992797	—	—	ID = Cflo_05684- -XP_394859.1_APIME;
scaffold859	CUFF.214386.0	5′ UTR	446051	446169	—	—	—	Parent = Cflo_05684- -XP_394859.1_APIME;
scaffold859	GLEAN	CDS	446008	446050	—	—	0	Parent = Cflo_05684- -XP_394859.1_APIME;
scaffold859	GLEAN	CDS	445823	445921	—	—	2	Parent = Cflo_05684- -XP_394859.1_APIME;
scaffold859	GLEAN	CDS	445583	445761	—	—	2	Parent = Cflo_05684- -XP_394859.1_APIME;
scaffold859	GLEAN	CDS	445110	445401	—	—	0	Parent = Cflo_05684- -XP_394859.1_APIME;
scaffold859	GLEAN	CDS	444447	444658	—	—	2	Parent = Cflo_05684- -XP_394859.1_APIME;
scaffold859	GLEAN	CDS	444197	444382	—	—	0	Parent = Cflo_05684- -XP_394859.1_APIME;
scaffold1661	GLEAN	mRNA	53519	59877	0.772684	+	—	ID = Cflo_12484- -XP_394081.2_APIME;
scaffold1661	CUFF.58661.0	5′ UTR	53519	53664	—	+	—	Parent = Cflo_12484- -XP_394081.2_APIME;
scaffold1661	GLEAN	CDS	53722	53867	—	+	0	Parent = Cflo_12484- -XP_394081.2_APIME;
scaffold1661	GLEAN	CDS	55325	55645	—	+	1	Parent = Cflo_12484- -XP_394081.2_APIME;
scaffold1661	GLEAN	CDS	56058	56438	—	+	1	Parent = Cflo_12484- -XP_394081.2_APIME;

TABLE 2. Annotation data resources

Data resource	Description	URL
Genome		
ASAP II	Database of splicing variants including tissue and cancer analysis (Kim et al. 2007)	http://bioinformatics.ucla.edu/ASAP2/
ASPicDB	Database of splicing pattern of human genes (Castrignano et al. 2008)	http://t.caspur.it/ASPicDB/
ASTD	Database containing alternative transcripts generated by either alternative splicing or alternative start or end points (Stamm et al. 2006)	http://www.ebi.ac.uk/astd/
dbSNP	A catalog of variation from the National Center for Biotechnology Information (Smigielski et al. 2000)	http://www.ncbi.nlm.nch.gov/projects/snp
Ensembl	Pipeline that includes prediction of genes, transcripts, and peptides (Flicek et al. 2008)	http://www.ensembl.org
FlyBase	Database of *Drosophila* genomes (Grumbling and Strelets 2006)	http://flybase.bio.indiana.edu/
GenBank	Database containing all publicly available DNA sequences (Benson et al. 2008)	http://www.ncbi.nlm.nih.gov/Genbank/
GOLD	Resource monitoring the worldwide genome projects (Liolios et al. 2008)	http://www.genomesonline.org/
NCBI tools	Repository of tools to perform analysis in several types of data: genes, proteins, and genomes (Wheeler et al. 2007)	http://www.ncbi.nlm.nih.gov/Tools/
OMIM	Database of human-inherited diseases and the genes causing them (Hamosh et al. 2002)	http://www.ncbi.nlm.nih.gov/omim/
RefSeq	Non-redundant database of annotated sequences (genomic DNA, transcripts, and proteins) (Pruitt et al. 2007)	http://www.ncbi.nlm.nih.gov/RefSeq/
SNPeffect	Database for the annotation of the effect of SNPs (Reumers et al. 2005)	http://snpeffect.vib.be/index.php
TAIR	Database containing genetic and molecular biology data for *Arabidopsis thaliana* (Swarbreck et al. 2008)	http://www.arabidopsis.org/
UCSC Genome Browser	Browser for displaying genomic data (Karolchik et al. 2008)	http://genome.ucsc.edu/
Vega	Repository of manually curated data for finished vertebrate genomes (Wilming et al. 2008)	http://vega.sanger.ac.uk
WormBase	Database containing genomic information for *Caenorhabditis elegans* and other nematodes (Rogers et al. 2008)	http://www.wormbase.org/

Continued

TABLE 2. *Continued*

Data resource	Description	URL
Proteomic/sequence		
A suite of tools to analyze post-translational modifications from the CBS	Predicting the attachment of chemical groups: phosphorylation (NetPhos [Blom et al. 1999]; NetPhosK [Blom et al. 2004]; NetPhosYeast [Ingrell et al. 2007]); *O*-linked glycosylation (NetOGlyc [Julenius et al. 2005]; YinOYang [Gupta and Brunak 2002]; DictyOGlyc [Gupta et al. 1999]); *N*-linked glycosylation (NetNGlyc); *C*-linked glycosylation (NetCGlyc [Julenius 2007]); glycation (NetGlycate [Johansen et al. 2006]); acetylation (NetAcet [Kiemer et al. 2005]); sulfation; and lipid attachment (LipoP [Juncker et al. 2003]). Tools for the indication of peptide cleavage: signal peptides (SignalP [Bendtsen et al. 2004]; LipoP [Juncker et al. 2003]; TatP [Bendtsen et al. 2005a, b]); propeptides (ProP [Duckert et al. 2004]); transit peptides (TargetP [Emanuelsson et al. 2007]; ChloroP [Emanuelsson et al. 1999]); viral polyprotein processing (NetCorona [Kiemer et al. 2004]; NetPicoRNA [Blom et al. 1996]); caspase cleavage and also protein sorting and subcellular localization; secretion (SecretomeP [Bendtsen et al. 2005a,b]); import into mitochondria and chloroplasts (ChloroP); and nuclear export (NetNES [La Cour et al. 2004])	http://www.cbs.dtu.dk/services/
CSA	Database containing information regarding catalytic residues, part manually curated, part by homology (Porter et al. 2004)	http://www.ebi.ac.uk/thornton-srv/databases/CSA/
FireDB/Firestar	Database containing residues with functional annotation (Lopez et al. 2007a) and a tool for predicting functional residues in unannotated sequences (Lopez et al. 2007b)	http://firedb.bioinfo.cnio.es/
Gene3D	Functional annotation database that searches similarities between unannotated proteins from whichever origin and CATH domains (Yeats et al. 2008)	http://gene3d.biochem.ucl.ac.uk/Gene3D/
Interpro	Consortium database that includes annotation from different database members (Mulder et al. 2007)	http://www.ebi.ac.uk/interpro/

TABLE 2. *Continued*

Data resource	Description	URL
iProClass	Integrative database for protein functional features (Wu et al. 2004)	http://pir.georgetown.edu/iproclass/
KEGG	Resource containing information regarding genes, functions, hierarchies, pathways, and ligands (Kanehisa et al. 2008)	http://www.genome.jp/kegg/
MEMSAT	Predicts the structure of all-helical transmembrane proteins and the location of their constituent helical elements within a membrane (Jones 2007)	http://bioinf.cs.ucl.ac.uk/memsat/
Panther	Database of functional assignments for genes and proteins (Thomas et al. 2003)	http://www.pantherdb.org/
Pfam	Database containing **multiple alignments** of protein domains and conserved regions (Finn et al. 2008)	http://www.sanger.ac.uk/Software/Pfam/
PIR	Databases and tools for genomic and proteomic studies (Wu et al. 2007)	http://pir.georgetown.edu/
PMut	Server aimed at the prediction of pathological mutations using neural networks (Ferrer-Costa et al. 2005a,b)	http://mmb.pcb.ub.es/PMut/
PRIDE	Repository for proteomics data, which allows users to submit, retrieve, and compare experimental data (Jones and Côté 2008)	http://www.ebi.ac.uk/pride/
Prints	Database of fingerprints characterizing protein families (Attwood 2002)	http://www.bioinf.manchester.ac.uk/dbbrowser/PRINTS/
ProDom	Database of protein domain families generated using SwissProt and TrEMBL sequences (Bru et al. 2005)	http://prodom.prabi.fr/
Prosite	Database of functional domains containing protein signatures (Hulo et al. 2006)	http://www.expasy.ch/prosite/
ProtoNet	Server that clusters proteins in order to predict structure and function (Kaplan et al. 2005)	http://www.protonet.cs.huji.ac.il/
PupaSuite	Web tool focused on the analysis of SNPs (Conde et al. 2006)	http://pupasuite.bioinfo.cipf.es/
SMART	Database of functional domains based on profiles obtained through hidden Markov models from homologous sequences (Schultz et al. 1998)	http://smart.embl-heidelberg.de/
Superfamily	Database of functional domain assignments (at the SCOP superfamily level) for completely sequenced organisms (Gough et al. 2001)	http://supfam.cs.bris.ac.uk/SUPERFAMILY/
TIGRFAMs	Database of protein families collated and annotated using HMMs (Haft et al. 2003)	http://www.tigr.org/TIGRFAMs/index.shtml

Continued

TABLE 2. *Continued*

Data resource	Description	URL
TMHMM	Prediction of transmembrane helices in proteins (Krogh et al. 2001)	http://www.cbs.dtu.dk/services/TMHMM/
UniprotKB/SwissProt	Database containing protein information features (The UniProt Consortium 2008)	www.ebi.ac.uk/swissprot/
UniprotKB/TrEMBL	Translated version of the EMBL database (The UniProt Consortium 2008)	http://www.ebi.ac.uk/TrEMBL/
Proteomic/structure		
CATH	Classification of protein domain structures mainly based on structural features (secondary structure, architecture, and topology) and homology clustering (Greene et al. 2007)	http://www.cathdb.info/
Genomic Threading Database	Proteome annotation from structure folding recognition (McGuffin et al. 2004)	http://bioinf.cs.ucl.ac.uk/GTD/
ModBase	Database of three-dimensional (3D) models built by homology modeling (Pieper et al. 2006)	http://modbase.compbio.ucsf.edu/
MoDEL	Database containing molecular dynamics trajectories and their analysis (Rueda et al. 2007)	http://mmb.pcb.ub.es/MODEL/
MSD	Collection, management, and distribution of data regarding macromolecular structures (Tagari et al. 2006)	http://www.ebi.ac.uk/msd/
PDBsum	Structural annotation of each 3D structure deposited in the protein Data Bank (Laskowski et al. 2005a)	http://www.ebi.ac.uk/thornton-srv/databases/pdbsum/
PISA	Tool to analyze PDB structures in order to predict the macromolecular interfaces and the quaternary state (Krissinel and Henrick 2007)	http://www.ebi.ac.uk/msd-srv/prot_int/pistart.html
Procognate	Database of cognate ligands for enzyme structures (Bashton et al. 2008)	http://www.ebi.ac.uk/thornton-srv/databases/procognate/
ProFunc	Identifies the likely biochemical function of a protein from its 3D structure (Laskowski et al. 2005b)	http://www.ebi.ac.uk/thornton-srv/databases/ProFunc/
RCSB PDB	Atlas of 3D protein structures into the PDB (Berman et al. 2002)	http://www.rcsb.org
SwissModel	Server aimed at the construction of homology models (Schwede et al. 2003)	http://swissmodel.expasy.org/SWISS-MODEL.html
SCOP	Structural classification of proteins based on evolutionary information and topology (Andreeva et al. 2004)	http://scop.mrc-lmb.cam.ac.uk/scop/

TABLE 2. *Continued*

Data resource	Description	URL
wwPDB	Repository aimed at maintaining a single protein Data Bank archive of macromolecular structural data (Berman et al. 2003)	http://www.wwpdb.org/index.html
Other		
ArrayExpress	Database containing curated expression profiles (Parkinson et al. 2007)	http://www.ebi.ac.uk/microarray-as/aew/
Babelomics	Integrated system for performing different analyses on gene function (Al-Shahrour et al. 2006)	http://babelomics.bioinfo.cipf.es/
Brenda	Database containing enzyme functional information such as K_m or substrates (Barthelmes et al. 2007)	http://www.brenda-enzymes.info/
ChEBI	Dictionary of small chemical compounds (Degtyarenko et al. 2008)	http://www.ebi.ac.uk/chebi/
GEPAS	Integrated system for performing different analyses on gene expression (Montaner et al. 2006)	http://gepas.bioinfo.cipf.es/
GSCAN	Server for the scanning of SNPs and QTLs in the genome (Valdar et al. 2006)	http://gscan.well.ox.ac.uk/
IntAct	Database containing molecular interaction data (Kerrien et al. 2007)	http://www.ebi.ac.uk/intact/
MACiE	Database of enzymatic reactions (Holliday et al. 2007)	http://www.ebi.ac.uk/thornton-srv/databases/MACiE/

Adapted from Reeves et al. 2009.

TABLE 3. Community-based genome annotation projects and software

Community	Tool	Website	Model organism
J. Craig Venter Institute	Manatee	http://manatee.sourceforge.net/	Prokaryotic organisms
Welcome Trust Sanger Institute	Artemis and ACT	http://www.sanger.ac.uk/Software/Artemis; http://www.sanger.ac.uk/Software/ACT	Invertebrate
Welcome Trust Sanger Institute	Zmap and Otterlace	http://www.acedb.org/Software/Downloads; ftp://ftp.sanger.ac.uk/pub2/jgrg	Vertebrate
The *Arabidopsis* Information Resource	Apollo	http://apollo.berkeleybop.org/current/index.html	*Arabidopsis*, plant
The Broad Institute of MIT	Argo	http://www.broadinstitute.org/annotation/argo/	

compared with the single-genome approach, leading to improved gene models, fewer missing elements, and more accuracy.

Reference-Based or Evidence-Based Approach

Reference-based approaches annotate genomes by training gene prediction program with a preexisting set of gene or genomic models for that species, which can be a set of EST references or protein domains. Taking advantage of recent advances in NGS technology, reference-based annotation strategies have been extended and enhanced. A **de novo RNA assembly** method by completely sequencing whole transcriptomes has been successfully implemented. This strategy theoretically allows exploring the whole transcriptome family in a rapid manner and building a complete reference of expressed RNA transcripts for use in annotating the genomic sequence assemblies. Some useful annotation tools and software are listed in Table 4.

Annotation Strategy and Workflow

Genome annotation is a multiple-step or -layer process. A general strategy that is widely used by the genome sequencing community is to (1) collect gene model evidence from an EST library or protein domain database or BLAST against known genomes; (2) build evidence-based gene models; (3) train the gene prediction program with the collected evidence; (4) predict gene models and other genome features; and (5) assign gene names and report annotation accuracy.

TRENDS AND ADVANCED PRACTICE

Genome annotation is a time-consuming and multiple-layer process. Neither individual ab initio nor more sophisticated computational tools is able to predict the gene model for a genome with accuracy >50%. The community-based effort has become more and more important for better genome annotation. The GENCODE Consortium and other collaborative teams (e.g., GMOD) have created a genome annotation framework for multiple model organisms, which have benefited from the continuous, in-depth annotation of expert curators and technology advances in both next-generation sequencing and software innovation.

The NGS technology has not only made genome sequencing faster and cheaper, but also allows researchers to generate transcriptomic evidence or genetic event–related evidence much more efficiently. Recent research has shown that the new technology has further advanced evidence-based annotation. Here, two technology advances in RNA-based de novo assembly that will promote evidence-based annotation are highlighted.

TABLE 4. Software tools for ab initio and evidence-based genome annotation

Name	Contributor	Web link	Main function
AUGUSTUS	Mario Stanke, Rasmus Steinkamp, Stephan Waack, and Burkhard Morgenstern	http://augustus.gobics.de/	Gene prediction in eukaryote model systems
CONTRAST	S.S. Gross, C.B. Do, M. Sirota, and S. Batzoglou	http://contra.stanford.edu/contrast	Gene prediction
EUGENE	T. Schiex, A. Moisan, and P. Rouzé. Computational Biology, Eds. O. Gascuel and M.-F. Sagot	http://eugene.toulouse.inra.fr/	Gene finder for eukaryotes
ExonHunter	Brona Brejova, Daniel G. Brown, Ming Li, and Tomas Vinar	http://compbio.fmph.uniba.sk/exonhunter/	Gene and exon prediction in multiple systems
FGENESH	Commercial package from Softberry	http://linux1.softberry.com/berry.phtml	Multi-model system gene prediction package
GeneID	E. Blanco, G. Parra, and R. Guigó	http://genome.crg.es/geneid.html	Ab initio gene model prediction in eukaryotes
GeneMark	M. Borodovsky and J. McIninch	http://exon.gatech.edu/	Ab initio gene model prediction in both prokaryotes and eukaryotes
GenScan	C. Berge	http://genes.mit.edu/GENSCAN.html	Ab initio gene model prediction in eukaryotes
GLIMMER	A.L. Delcher, D. Harmon, S. Kasif, O. White, and S.L. Salzberg	http://www.cbcb.umd.edu/software/glimmer/	Ab initio gene model prediction in prokaryotes
Gnomon	NCBI	http://www.ncbi.nlm.nih.gov/projects/genome/guide/gnomon.shtml	Gene prediction in eukaryote model systems
mGene	G. Rätsch	http://fml.tuebingen.mpg.de/raetsch/suppl/mgene	Gene prediction
mSplicer	Gunnar Rätsch, Sören Sonnenburg, Jagan Srinivasan, Hanh Witte, Klaus-Robert Müller, Ralf Sommer, and Bernhard Schölkopf	http://www.fml.tuebingen.mpg.de/raetsch/suppl/msplicer	Gene splicing prediction in worm genome
NNPP	M.G. Reese	http://www.fruitfly.org/seq_tools/promoter.html	Promoter prediction
NNSPLICE	M.G. Reese	http://www.fruitfly.org/seq_tools/splice.html	Splicing site prediction
ORF FINDER	T. Tatusov and R. Tatusov	http://www.ncbi.nlm.nih.gov/gorf/gorf.html	Finding open read frame

Continued

TABLE 4. *Continued*

Name	Contributor	Web link	Main function
SGP	G. Parra, P. Agarwal, J.F. Abril, T. Wiehe, J.W. Fickett, and R. Guigó	http://jakob.genetik.uni-koeln.de/bioinformatik/software/	Cross-reference gene finding
SLAM	Simon Cawley, Lior Pachter, and Marina Alex	http://bio.math.berkeley.edu/slam/	Cross-reference gene finding
SNAP	I. Korf	http://homepage.mac.com/iankorf/	Ab initio gene prediction for multiple model systems
TWINSCAN/N-SCAN	I. Korf, P. Flicek, D. Duan, and M. R. Brent	http://mblab.wustl.edu/software/twinscan/	Cross-reference gene finding

De Novo Transcriptomic Assembly with Velvet and Oases

De novo RNA assembly uses very short RNA sequence tags by combining the Velvet and Oases software packages. The technology allows researchers to study a transcriptome without a complete genome assembly. The same RNA **short reads** from the study can be used for multiple purposes: for example, to assemble the whole transcriptome or to study transcriptomic events such as differential gene expression. In contrast to the traditional approach of obtaining the genome assembly before genome-wide transcriptome study, this allows genome-wide gene expression study for a new genome in a much faster and cheaper way. Moreover, the de novo RNA assembly will further help the genome assembly and annotation.

De Novo Transcriptomic Assembly with Trinity

The Trinity technology requires much longer short reads, usually 76 bp from both ends. This will increase the cost a little, but could improve the assembly in terms of transcriptome quality.

REFERENCES

Al-Shahrour F, Minguez P, Tárraga J, Montaner D, Alloza E, Vaquerizas JM, Conde L, Blaschke C, Vera J, Dopazo J. 2006. BABELOMICS: A systems biology perspective in the functional annotation of genome-scale experiments. *Nucleic Acids Res* **34:** W472–W476.

Andreeva A, Howorth D, Brenner SE, Hubbard TJP, Chothia C, Murzin AG. 2004. SCOP database in 2004: Refinements integrate structure and sequence family data. *Nucleic Acids Res* **32:** D226–D229.

Attwood TK. 2002. The PRINTS database: A resource for identification of protein families. *Brief Bioinform* **3:** 252–263.

Barthelmes J, Ebeling C, Chang A, Schomburg I, Schomburg D. 2007. BRENDA, AMENDA and FRENDA: The enzyme information system in 2007. *Nucleic Acids Res* **35:** D511–D514.

Bashton M, Nobeli I, Thornton JM. 2008. PROCOGNATE: A cognate ligand domain mapping for enzymes. *Nucleic Acids Res* **36:** D618–D622.

Bendtsen JD, Nielsen H, von Heijne G, Brunak S. 2004. Improved prediction of signal peptides: SIGNALP 3.0. *J Mol Biol* **340:** 783–795.

Bendtsen JD, Kiemer L, Fausbøll A, Brunak S. 2005a. Non-classical protein secretion in bacteria. *BMC Microbiol* **5:** 58.

Bendtsen JD, Nielsen H, Widdick D, Palmer T, Brunak S. 2005b. Prediction of twin-arginine signal peptides. *BMC Bioinform* **6:** 167.

Benson DA, Karsch-Mizrachi I, Lipman DJ, Ostell J, Wheeler DL. 2008. GenBank. *Nucleic Acids Res* **36:** D25–D30.

Berman HM, Battistuz T, Bhat TN, Bluhm WF, Bourne PE, Burkhardt K, Feng Z, Gilliland GL, Iype L, Jain S, et al. 2002. The Protein Data Bank. *Acta Crystallogr D Biol Crystallogr* **58:** 899–907.

Berman H, Henrick K, Nakamura H. 2003. Announcing the worldwide Protein Data Bank. *Nat Struct Biol* **10:** 980.

Blom N, Hansen J, Blaas D, Brunak S. 1996. Cleavage site analysis in picornaviral polyproteins: Discovering cellular targets by neural networks. *Protein Sci* **5:** 2203–2216.

Blom N, Gammeltoft S, Brunak S. 1999. Sequence and structure-based prediction of eukaryotic protein phosphorylation sites. *J Mol Biol* **294:** 1351–1362.

Blom N, Sicheritz-Pontén T, Gupta R, Gammeltoft S, Brunak S. 2004. Prediction of post-translational glycosylation and phosphorylation of proteins from the amino acid sequence. *Proteomics* **4:** 1633–1649.

Bru C, Courcelle E, Carrère S, Beausse Y, Dalmar S, Kahn D. 2005. The PRODOM database of protein domain families: More emphasis on 3D. *Nucleic Acids Res* **33:** D212–D215.

Castrignano T, D'Antonio M, Anselmo A, Carrabino D, D'Onorio De Meo A, D'Erchia AM, Licciulli F, Mangiulli M, Mignone F, Pavesi G, et al. 2008. ASPICDB: A database resource for alternative splicing analysis. *Bioinformatics* **24:** 1300–1304.

Conde L, Vaquerizas JM, Dopazo H, Arbiza L, Reumers J, Rousseau F, Schymkowitz J, Dopazo J. 2006. PUPASUITE: Finding functional single nucleotide polymorphisms for large-scale genotyping purposes. *Nucleic Acids Res* **34:** W621–W625.

Degtyarenko K, de Matos P, Ennis M, Hastings J, Zbinden M, McNaught A, Alcántara R, Darsow M, Guedj M, Ashburner M. 2008. CHEBI: A database and ontology for chemical entities of biological interest. *Nucleic Acids Res* **36:** D344–D350.

Duckert P, Brunak S, Blom N. 2004. Prediction of proprotein convertase cleavage sites. *Protein Eng Des Sel* **17:** 107–112.

Emanuelsson O, Nielsen H, von Heijne G. 1999. CHLOROP, a neural network-based method for predicting chloroplast transit peptides and their cleavage sites. *Protein Sci* **8:** 978–984.

Emanuelsson O, Brunak S, von Heijne G, Nielsen H. 2007. Locating proteins in the cell using TARGETP, SIGNALP and related tools. *Nat Protoc* **2:** 953–971.

The ENCODE Project Consortium. 2011. A user's guide to the Encyclopedia of DNA Elements (ENCODE). *PLoS Biol* **9:** 1–21.

Ferrer-Costa C, Gelpí JL, Zamakola L, Parraga I, de la Cruz X, Orozco M. 2005a. PMUT: A web-based tool for the annotation of pathological mutations on proteins. *Bioinformatics* **21:** 3176–3178.

Ferrer-Costa C, Orozco M, de la Cruz X. 2005b. Use of bioinformatics tools for the annotation of disease-associated mutations in animal models. *Proteins* **61:** 878–887.

Finn RD, Tate J, Mistry J, Coggill PC, Sammut SJ, Hotz H-R, Ceric G, Forslund K, Eddy SR, Sonnhammer EL, Bateman A. 2008. The PFAM protein families database. *Nucleic Acids Res* **36:** D281–D288.

Flicek P, Aken BL, Beal K, Ballester B, Caccamo M, Chen Y, Clarke L, Coates G, Cunningham F, Cutts T, et al. 2008. ENSEMBL 2008. *Nucleic Acids Res* **36:** D707–D714.

Gough J, Karplus K, Hughey R, Chothia C. 2001. Assignment of homology to genome sequences using a library of hidden Markov models that represent all proteins of known structure. *J Mol Biol* **313:** 903–919.

Greene LH, Lewis TE, Addou S, Cuff A, Dallman T, Dibley M, Redfern O, Pearl F, Nambudiry R, Reid A, et al. 2007. The CATH domain structure database: New protocols and classification levels give a more comprehensive resource for exploring evolution. *Nucleic Acids Res* **35:** D291–D297.

Grumbling G, Strelets V. 2006. FLYBASE: Anatomical data, images and queries. *Nucleic Acids Res* **34:** D484–D488.

Gupta R, Brunak S. 2002. Prediction of glycosylation across the human proteome and the correlation to protein function. *Pac Symp Biocomput* **7:** 310–322.

Gupta R, Jung E, Gooley AA, Williams KL, Brunak S, Hansen J. 1999. Scanning the available *Dictyostelium discoideum* proteome for *O*-linked GlcNAc glycosylation sites using neural networks. *Glycobiology* **9:** 1009–1022.

Haft DH, Selengut JD, White O. 2003. The TIGRFAMs database of protein families. *Nucleic Acids Res* **31:** 371–373.

Hamosh A, Scott AF, Amberger J, Bocchini C, Valle D, McKusick VA. 2002. Online Mendelian inheritance in man (OMIM), a knowledgebase of human genes and genetic disorders. *Nucleic Acids Res* **30:** 52–55.

Holliday GL, Almonacid DE, Bartlett GJ, O'Boyle NM, Torrance JW, Murray-Rust P, Mitchell JBO, Thornton JM. 2007. MACIE (mechanism, annotation and classification in enzymes): Novel tools for searching catalytic mechanisms. *Nucleic Acids Res* **35:** D515–D520.

Hulo N, Bairoch A, Bulliard V, Cerutti L, De Castro E, Langendijk-Genevaux PS, Pagni M, Sigrist CJ. 2006. The PROSITE database. *Nucleic Acids Res* **34:** D227–D230.

Ingrell CR, Miller ML, Jensen ON, Blom N. 2007. NETPHOSYEAST: Prediction of protein phosphorylation sites in yeast. *Bioinformatics* **23:** 895–897.

Johansen MB, Kiemer L, Brunak S. 2006. Analysis and prediction of mammalian protein glycation. *Glycobiology* **16:** 844–853.

Jones DT. 2007. Improving the accuracy of transmembrane protein topology prediction using evolutionary information. *Bioinformatics* **23:** 538–544.

Jones P, Côté R. 2008. The PRIDE proteomics identifications database: Data submission, query, and dataset comparison. *Methods Mol Biol* **484:** 287–303.

Julenius K. 2007. NETCGLYC 1.0: Prediction of mammalian C-mannosylation sites. *Glycobiology* **17:** 868–876.

Julenius K, Mølgaard A, Gupta R, Brunak S. 2005. Prediction, conservation analysis, and structural characterization of mammalian mucin-type *O*-glycosylation sites. *Glycobiology* **15:** 153–164.

Juncker AS, Willenbrock H, Von Heijne G, Brunak S, Nielsen H, Krogh A. 2003. Prediction of lipoprotein signal peptides in Gram-negative bacteria. *Protein Sci* **12:** 1652–1662.

Kanehisa M, Araki M, Goto S, Hattori M, Hirakawa M, Itoh M, Katayama T, Kawashima S, Okuda S, Tokimatsu T, Yamanishi Y. 2008. KEGG for linking genomes to life and the environment. *Nucleic Acids Res* **36:** D480–D484.

Kaplan N, Sasson O, Inbar U, Friedlich M, Fromer M, Fleischer H, Portugaly E, Linial N, Linial M. 2005. PROTONET 4.0: A hierarchical classification of one million protein sequences. *Nucleic Acids Res* **33:** D216–D218.

Karolchik D, Kuhn RM, Baertsch R, Barber GP, Clawson H, Diekhans M, Giardine B, Harte RA, Hinrichs AS, Hsu F, et al. 2008. The UCSC Genome browser database: 2008 update. *Nucleic Acids Res* **36:** D773–D779.

Kerrien S, Alam-Faruque Y, Aranda B, Bancarz I, Bridge A, Derow C, Dimmer E, Feuermann M, Friedrichsen A, Huntley R, et al. 2007. INTACT—Open source resource for molecular interaction data. *Nucleic Acids Res* **35:** D561–D565.

Kiemer L, Lund O, Brunak S, Blom N. 2004. Coronavirus 3CLpro proteinase cleavage sites: Possible relevance to SARS virus pathology. *BMC Bioinform* **5:** 72.

Kiemer L, Bendtsen JD, Blom N. 2005. NETACET: Prediction of N-terminal acetylation sites. *Bioinformatics* **21:** 1269–1270.

Kim N, Alekseyenko AV, Roy M, Lee C. 2007. The ASAP II database: Analysis and comparative genomics of alternative splicing in 15 animal species. *Nucleic Acids Res* **35:** D93–D98.

Krissinel E, Henrick K. 2007. Inference of macromolecular assemblies from crystalline state. *J Mol Biol* **372:** 774–797.

Krogh A, Larsson B, von Heijne G, Sonnhammer EL. 2001. Predicting transmembrane protein topology with a hidden Markov model: Application to complete genomes. *J Mol Biol* **305:** 567–580.

La Cour T, Kiemer L, Mølgaard A, Gupta R, Skriver K, Brunak S. 2004. Analysis and prediction of leucine-rich nuclear export signals. *Protein Eng Des Sel* **17:** 527–536.

Laskowski RA, Chistyakov VV, Thornton JM. 2005a. PDBSUM more: New summaries and analyses of the known 3D structures of proteins and nucleic acids. *Nucleic Acids Res* **33:** D266–D268.

Laskowski RA, Watson JD, Thornton JM. 2005b. PROFUNC: A server for predicting protein function from 3D structure. *Nucleic Acids Res* **33:** W89–W93.

Liolios K, Mavromatis K, Tavernarakis N, Kyrpides NC. 2008. The genomes on line database (GOLD) in 2007: Status of genomic and metagenomic projects and their associated metadata. *Nucleic Acids Res* **36:** D475–D479.

Lopez G, Valencia A, Tress M. 2007a. FIREDB—A database of functionally important residues from proteins of known structure. *Nucleic Acids Res* **35:** D219–D223.

Lopez G, Valencia A, Tress ML. 2007b. FIRESTAR—Prediction of functionally important residues using structural templates and alignment reliability. *Nucleic Acids Res* **35:** W573–W577.

McGuffin LJ, Street SA, Bryson K, Sørensen S-A, Jones DT. 2004. The genomic threading database: A comprehensive resource for structural annotations of the genomes from key organisms. *Nucleic Acids Res* **32:** D196–D199.

Montaner D, Tárraga J, Huerta-Cepas J, Burguet J, Vaquerizas JM, Conde L, Minguez P, Vera J, Mukherjee S, Valls J, et al. 2006. Next station in microarray data analysis: GEPAS. *Nucleic Acids Res* **34:** W486–W491.

Mulder NJ, Apweiler R, Attwood TK, Bairoch A, Bateman A, Binns D, Bork P, Buillard V, Cerutti L, Copley R, et al. 2007. New developments in the INTERPRO database. *Nucleic Acids Res* **35:** D224–D228.

Parkinson H, Kapushesky M, Shojatalab M, Abeygunawardena N, Coulson R, Farne A, Holloway E, Kolesnykov N, Lilja P, Lukk M, et al. 2007. ARRAYEXPRESS—A public database of microarray experiments and gene expression profiles. *Nucleic Acids Res* **35:** D747–D750.

Pieper U, Eswar N, Davis FP, Braberg H, Madhusudhan MS, Rossi A, Marti-Renom M, Karchin R, Webb BM, Eramian D, et al. 2006. MODBASE: A database of annotated comparative protein structure models and associated resources. *Nucleic Acids Res* **34:** D291–D295.

Porter CT, Bartlett GJ, Thornton JM. 2004. The catalytic site atlas: A resource of catalytic sites and residues identified in enzymes using structural data. *Nucleic Acids Res* **32:** D129–D133.

Pruitt KD, Tatusova T, Maglott DR. 2007. NCBI reference sequences (REFSEQ): A curated non-redundant sequence database of genomes, transcripts and proteins. *Nucleic Acids Res* **35:** D61–D65.

Reumers J, Schymkowitz J, Ferkinghoff-Borg J, Stricher F, Serrano L, Rousseau F. 2005. SNPEFFECT: A database mapping molecular phenotypic effects of human non-synonymous coding SNPs. *Nucleic Acids Res* **33:** D527–D532.

Rogers A, Antoshechkin I, Bieri T, Blasiar D, Bastiani C, Canaran P, Chan J, Chen WJ, Davis P, Fernandes J, et al. 2008. WORMBASE 2007. *Nucleic Acids Res* **36:** D612–D617.

Rueda M, Ferrer-Costa C, Meyer T, Pérez A, Camps J, Hospital A, Gelpí JL, Orozco M. 2007. A consensus view of protein dynamics. *Proc Natl Acad Sci* **104:** 796–801.

Schultz J, Milpetz F, Bork P, Ponting CP. 1998. SMART, a simple modular architecture research tool: identification of signaling domains. *Proc Natl Acad Sci* **95:** 5857–5864.

Schwede T, Kopp J, Guex N, Peitsch MC. 2003. SWISS-MODEL: An automated protein homology-modeling server. *Nucleic Acids Res* **31:** 3381–3385.

Smigielski EM, Sirotkin K, Ward M, Sherry ST. 2000. DBSNP: A database of single nucleotide polymorphisms. *Nucleic Acids Res* **28:** 352–355.

Stamm S, Riethoven JJ, Le TV, Gopalakrishnan C, Kumanduri V, Tang Y, Barbosa-Morais NL, Thanaraj TA. 2006. ASD: A bioinformatics resource on alternative splicing. *Nucleic Acids Res* **34:** D46–D55.

Stein L. 2001. Genome annotation: From sequence to biology. *Nat Rev Genet* **2:** 493–503.

Swarbreck D, Wilks C, Lamesch P, Berardini TZ, Garcia-Hernandez M, Foerster H, Li D, Meyer T, Muller R, Ploetz L, et al. 2008. The *Arabidopsis* information resource (TAIR): Gene structure and function annotation. *Nucleic Acids Res* **36:** D1009–D1014.

Tagari M, Tate J, Swaminathan GJ, Newman R, Naim A, Vranken W, Kapopoulou A, Hussain A, Fillon J, Henrick K, Velankar S. 2006. E-MSD: Improving data deposition and structure quality. *Nucleic Acids Res* **34:** D287–D290.

Thomas PD, Campbell MJ, Kejariwal A, Mi H, Karlak B, Daverman R, Diemer K, Muruganujan A, Narechania A. 2003. PANTHER: A library of protein families and subfamilies indexed by function. *Genome Res* **13:** 2129–2141.

The UniProt Consortium. 2008. The Universal Protein Resource (UniProt). *Nucleic Acids Res* **36:** D190–D195.

Valdar W, Solberg LC, Gauguier D, Burnett S, Klenerman P, Cookson WO, Taylor MS, Rawlins JN, Mott R, Flint J. 2006. Genome-wide genetic association of complex traits in heterogeneous stock mice. *Nat Genet* **38:** 879–887.

Wheeler DL, Barrett T, Benson DA, Bryant SH, Canese K, Chetvernin V, Church DM, DiCuccio M, Edgar R, Federhen S, et al. 2007. Database resources of the National Center for Biotechnology Information. *Nucleic Acids Res* **35:** D5–D12.

Wilming LG, Gilbert JGR, Howe K, Trevanion S, Hubbard T, Harrow JL. 2008. The Vertebrate Genome Annotation (VEGA) database. *Nucleic Acids Res* **36:** D753–D760.

Wu CH, Huang H, Nikolskaya A, Hu Z, Barker WC. 2004. The IPROCLASS integrated database for protein functional analysis. *Comput Biol Chem* **28:** 87–96.

Wu Q, Gaddis SS, MacLeod MC, Walborg EF, Thames HD, DiGiovanni J, Vasquez KM. 2007. High-affinity triplex-forming oligonucleotide target sequences in mammalian genomes. *Mol Carcinog* **46:** 15–23.

Yeats C, Lees J, Reid A, Kellam P, Martin N, Liu X, Orengo C. 2008. GENE3D: Comprehensive structural and functional annotation of genomes. *Nucleic Acids Res* **36:** D414–D418.

WWW RESOURCE

http://www.ensemblgenomes.org/ Ensembl Genomes annotation system

8

Using NGS to Detect Sequence Variants

Jinhua Wang, Zuojian Tang, and Stuart M. Brown

In general, predicting single-nucleotide variants (SNVs) involves the following steps. Each step can be altered/optimized to achieve the best performance based on the sequencing quality, **coverage**, and experimental design. We cover the following steps in detail:

1. quality control, error models (overlapping reads, **paired-end reads**, high error at end, trimming issues, quality trimming, etc.)
2. BAM processing and analysis tools
3. single nucleic variation detection: SNV call from **RNA-seq**, DNA exome, whole genome, SNV call without **reference genome** sequences (examples, tools)
4. SNV evaluation and annotation, widespread RNA and DNA sequence differences in the human transcriptome (tools, pipelines)
5. cancer-specific variant discovery (tools, examples)

INTRODUCTION

The detection of **sequence variants** is one of the most common applications of **next-generation sequencing** (NGS). This approach is widely used in both basic research, such as studies of genetic diversity and evolution, as well as direct clinical applications, such as cancer genomics and the discovery of mutations causing rare genetic diseases. Sequence variants occur in many different forms including single base pair changes, generally known as single-nucleotide polymorphisms (SNPs); insertion and deletion (**indel**) events (which may be as small as a single base pair or as large as millions of base pairs); translocations; inversions; and copy number variations (CNVs) in regions of sequence repeats. SNPs and single base insertion/deletion events are often grouped together as single-nucleotide variants (SNVs) because many software packages detect both at the same time, and they can have similar impacts on gene function. Translocations, inversions, large deletions (or insertions), and copy number changes are often grouped together as structural variants (SVs) because they all have the potential for large changes in gene function including the loss of multiple genes, amplification of gene expression, and the creation of novel gene fusions with potentially dominant effects. Each of these types of sequence variation can be detected by NGS, but each requires optimized experimental designs and software. Important biological discoveries have been published using NGS to detect each of these types of sequence variants, but because methods are still very much under development, benchmarks and best practices have not yet been established.

Detection and study of SNPs was well established before NGS machines were developed. The GenBank Database of Single Nucleotide Polymorphisms (dbSNP) was created in 1998 as a central public repository of genetic variation for humans and model organisms (Sherry et al. 1999, 2001). As of dbSNP build 132 (September 2010), there are 38 million human SNPs mapped to unique locations on the human genome **reference sequence**. This gives an approximate frequency of one SNP per 1200 bases for any person versus the reference genome, or between any two randomly chosen (unrelated) people. As a general rule, the variants in dbSNP are present at moderately high frequency (>5%) in at least some segment of the human population, and many of the SNPs have been genotyped across several ethnically different sample populations. The variants in dbSNP were used by several commercial vendors (including Illumina, Affymetrix, and Agilent) to create **microarray**-based genome-wide SNP genotyping assays, which became the basis for the field of Genome Wide Association (GWAS).

SNPs are discovered by sequencing, aligning new experimentally determined sequences to the reference genome, and identifying the differences. The critical aspect of SNP discovery is to prove that a new variant found in a sequenced fragment is genuinely present in the genome of the cells that were sampled, and not an artifact of **cloning** or sequencing. Most SNP discovery pipelines require that multiple non-identical reads contain the same variant. This is complicated by the fact that humans (and most other eukaryotes) are **diploid**. Therefore, a variant may be **heterozygous** and thus present in only 50% of reads covering a specific locus. Another problem is that all sequencing technologies produce errors—incorrectly called bases (see Chapter 1). In some cases, these errors are systematic rather than purely random, and thus a specific base may be miscalled in the same way in multiple reads. In some (but not all) cases, sequences determined on the opposite strand may escape a source of systematic error. The probability of sequencing error for each base is captured in the sequence quality score. Quality scores for **Sanger sequencing** on ABI fluorescent sequencers (slab gels and capillary machines) are very accurately estimated by the Phred program (see Chapter 2). Quality scores for sequences produced by NGS machines are estimated by software developed by the manufacturers. For **Illumina sequencers**, the quality score represents the ratio of the strongest fluorescent signal for a base to the next strongest signal. Some studies have claimed that these quality scores do not accurately reflect the true probability of error for each base (Bravo and Irizarry 2010). In particular, the position of a base within a **sequence read** (at the beginning, middle, or end) can have a very large effect on its probability of error, which is not adequately reflected in the current quality scores reported by NGS machines. The immediate sequence context (adjacent bases) can also have an effect on error. Certain dinucleotides are more often associated with errors on each sequencing platform. Quality score recalibration is now a feature of several **variant detection** methods such as GATK (McKenna et al. 2010), which reprocess quality scores after **alignment**, taking into account read position (machine cycle), dinucleotide context, and a baseline expected error rate (as calculated from loci expected to have no SNPs).

All of the existing software tools for SNP discovery using NGS data are postprocessors for reads aligned to a reference genome (see Chapter 4), or they detect SNPs during the alignment process. Some tools require alignments produced by a specific software package, whereas others are able to read alignments in standard data formats, such as **FASTQ** and **SAM/BAM** (see Chapter 3). These tools identify SNPs as differences between sequence reads and the reference genome, then use a variety of filters based on depth of sequencing coverage, frequency of the variant base among the reads, and quality of the base calls both at the variant position and nearby, as well as other clues from the local sequence region such as the presence of repeated sequences and reads with insertion/deletion variants. A local cluster of SNPs may be due to one or more low-quality reads, or an alignment error. SNPs near indels are filtered because alignment software has difficulty with indels, incorrectly placing gaps, so that the short end of a read after an indel is often incorrectly aligned. Quality scores for alignments are also important inputs for **variant detection** (Li et al. 2008). These filters have been developed by empirical observation of false positives. Extreme stringency is needed, because even one false positive per million bases of genome sequence would identify thousands of false mutations in the clinical sequencing of a single patient.

Low coverage is an obvious problem in the presence of sequencing errors (typically 0.5%–1% in Illumina sequencing) because one variant found in a single read cannot be distinguished from a random error, but a variant found in 40% of 10 or 20 reads is unlikely to be due to base-calling error. Low coverage in a specific region could be an indicator of an alignment problem due to local sequence anomaly such as copy number variation, structural rearrangement, or high level of sequence variation in the sample compared with the reference genome. The major histocompatibility complex locus is often cited as a problematic region for alignment and SNP calling, with both structural rearrangements and local sequence variation. Very high coverage (much higher than the rest of the genome) is an indicator of a possible PCR amplification artifact or an alignment error (such as a repeated sequence). It is widely recommended to remove duplicate reads because they create bias in coverage and **allele** frequency that is not an accurate reflection of the sample. The ideal coverage is generally found in the range of 20× to 50×. In general, a large number of sequence differences between the sample and the reference genome makes alignment more difficult, which, in turn, makes variant detection less accurate.

Rigorous study of SNP detection in large genome sequencing projects such as the 1000 Genomes and the Cancer Genome Atlas has revealed unacceptably high rates of incorrect SNP calls using "ab initio" methods that rely on the data within a single sample or sequencing lane. Improved SNP calling methods have been developed using a Bayesian framework that estimates the probability of each possible genotype in a sample taking into account both data from the sequence reads as well as prior data regarding that specific locus on the reference genome (Li et al. 2009). Sources of prior data include dbSNP, HapMap, and SeattleSNPs, as well as sequence reads from other samples collected in the same study or at the same sequencing center. Family and population structure are also valuable sources of covariate information. It becomes much easier to call a SNP that is present in 12% of reads from one patient at a particular genomic locus if it is also known that a sequence variant of that type exists in the relevant population at 18% allele frequency. This method has been further extended to take into account known haplotype frequencies across genomic intervals (Altshuler et al. 2010). However, prior information is not very helpful in calling very rare or novel SNPs, and these methods are only possible when a large whole-genome database of genotypes is available—which is only true for humans at this point.

CANCER-SPECIFIC VARIANT DISCOVERY

Predicting SNVs from next-generation sequencing of tumors is extremely challenging. There are various tools existing for SNV discovery from NGS data, but very few are specifically suited to work with data from tumors where altered ploidy and tumor cellularity impact the statistical expectation of SNV discovery. Cancer tumor samples present special challenges for sequence variant detection. Most somatic mutations are expected to occur on just one chromosome; thus they will be heterozygous. In addition, tumor biopsy samples contain a mixture of tumor and normal somatic cells. Therefore, the expected allele frequency of a SNV may be substantially less than 50%, which makes true mutations difficult to distinguish from sequencing errors unless the coverage is extremely deep.

Cancer is a disease of genetic alterations. In particular, SNVs present as either germline or somatic point mutations are essential causes of tumorigenesis and cellular proliferation in many human cancer types. The discovery of germline mutations established important gene functions in cancer; however, the contribution of single germline alleles to the population burden of cancer is relatively low. In contrast, determination of tumorigenic mechanisms has focused on somatic mutations. The somatic mutational landscape of cancer has, to date, largely been derived from small-scale or targeted approaches, leading to the discovery of genes affected by somatic mutations in many diverse cancer types. More comprehensive studies using Sanger-based **exon** resequencing suggest that the mutational landscape will be characterized by relative handfuls of frequently mutated genes and a long tail of infrequent somatic mutations in many genes. These mutations may also be characterized as "drivers," which directly contribute to cellular proliferation or other pathological processes, and "passengers," which are simply a consequence of dysfunctional DNA replication and DNA repair systems in precancerous or cancerous cells.

Cancer sequencing projects often compare the genome of a tumor sample (biopsy) with normal tissue from the same patient, with the goal to discover the specific somatic mutations that led to carcinogenesis or malignant transformation. These mutations may be clinically important as indicators for specific drugs or other therapeutic interventions. In this scenario, the stringent filters established to prevent false-positive SNV calls often create a new problem with false negatives. Mutations found in the tumor sample are screened against those found in the normal sample to find tumor-specific mutations. In many cases, these tumor-specific mutations are actually false negatives—where the mutation was present in the normal sample but was filtered out by overly stringent thresholds. A pragmatic solution to this problem is to use different SNV calling thresholds for tumor and normal samples, but this is very inelegant.

MAQ

Mapping and Assemblies with Qualities (MAQ) was the first widely used tool for SNV detection in very-high-throughput NGS (from Illumina GA and ABI SOLiD machines). MAQ is designed to map **short reads** to a reference genome, allowing for mismatches, and using quality scores to define SNVs (Li et al. 2008). The authors of this software have moved on to other methods, thus it is no longer being updated (the most recent version of MAQ on SourceForge is from 2008: http://sourceforge.net/projects/maq/files). Because the maximum read length MAQ can align is 63 bp, it cannot work with data from the Illumina HiSeq and other newer machines. MAQ can map ~1 million reads to the human genome in 10 CPU hours. Note that MAQ runs on a single processor; it has not been parallelized in any way. A quality score is calculated for the alignment of each read to the genome (error probability of the alignment). It calls the **consensus** genotypes, including homozygous and heterozygous polymorphisms, with a Phred probabilistic quality assigned to each base. MAQ can also use paired-end data to detect both short and long indels.

MAQ performs ungapped alignments. It can map reads to the reference genome with up to three mismatches. It finds all mismatched bases and calls SNPs based on a statistical model, which maximizes the posterior probability and calculates a Phred quality at each position along the consensus. Heterozygotes are also called in this process.

MAQ is relatively easy to use but requires some UNIX skills. To install it requires a GNU make command: `[ ./configure; make; make install]`; if this looks unfamiliar, get help. Most of the functions of MAQ are built into a single `easyrun` command in a Perl script: `maq.pl easyrun -d outdir ref.fasta reads.fastq`, where *ref.fasta* is the reference sequence in **FASTA format** and *reads.fasta* is the NGS data in FASTQ format. This produces an overall consensus alignment with quality scores and a list of SNPs. In addition, a "**pileup**" alignment format is also available that produces a line of output for each base in the reference sequence. Each line consists of chromosome, position, reference base, depth, and the bases on reads that cover this position (bases identical to the reference are shown as a comma or dot and read bases different from the reference in letters), base qualities, and mapping qualities. This pileup format has been used as input for some other SNP calling and genome visualization software.

The SNP output file has its own format. Each line consists of chromosome, position, reference base, consensus base, Phred-like consensus quality, read depth, the average number of hits of reads covering this position, the highest mapping quality of the reads covering the position, the minimum consensus quality in the 3-bp flanking regions at each side of the site (6 bp in total), the second best call, log-likelihood ratio of the second best and the third best call, and the third best call.

MAQ has relatively few options for tweaking the SNP calls, but an additional command, `maq.pl SNPfilter`, does provide some parameters:

`d` *INT [3]* Minimum read depth required to call a SNP

`-D` *INT [256]* Maximum read depth allowed to call a SNP (<255; otherwise ignored)

`-Q` *INT [40]* Required mapping quality of reads covering the SNP

`-q` *INT [20]* Minimum consensus quality

`-n` *INT [20]* Minimum adjacent consensus quality

`-w` *INT [3]* Size of the window around the potential indels. SNPs that are close to indels will be suppressed.

BWA and SAMtools

BWA and SAMtools are more powerful tools that have largely replaced MAQ for alignment of short NGS reads to a reference genome and detection of SNVs. BWA is the Burrows–Wheeler Alignment Tool, which aligns short reads to a reference genome based on the **Burrows–Wheeler transformation** (Li and Durbin 2009). BWA is currently "state of the art" for NGS alignment; its speed is far superior to earlier methods such as **Smith–Waterman, BLAST**, or MAQ. BWA can align 2 million short NGS reads to the human genome in ~20 min on one CPU (using ~3 GB of

RAM) with flexible tolerance for mismatches and indels. BWA gains its speed from a clever indexing scheme (Burrows–Wheeler transformation) that compresses the reference genome into a suffix array. See Chapter 4 for a detailed discussion of the **alignment algorithm** used by BWA.

BWA does not call SNVs, but it produces alignment output in the SAM format (Sequence Alignment/Map), which can be processed by several variant detection tools. The SAM format has been widely adopted for the storage and exchange of NGS data aligned to a reference genome. The SAM format contains an optional header and a line with 11 mandatory fields that provide the sequence read, quality, position on the reference genome, and differences between the read and the reference (SNVs) defined in a shorthand code called CIGAR. SAM-formatted files can be compressed into an indexed binary form called BAM, which is more efficient in disk space use and in the speed by which the data can be accessed by software (see Table 1).

SAMtools is a SNV finder that works with SAM (and BAM)–formatted alignments of NGS data. It has a very large set of customizable parameters (http://samtools.sourceforge.net/samtools.shtml). SAMtools requires a reference genome, indexed in its own format, and position-sorted alignment files of NGS data. SAMtools produces output in the relatively new Variant Call Format (VCF; http://vcftools.sourceforge.net/specs.html) (see Table 2).

SAMtools can filter SNPs on coverage (read depth), number of reads containing reference and alternate bases, average quality score of reference and alternate bases, mapping quality of the reads, and the existence of nearby indels in the reads (which tend to create false positives) using a measure called Base Alignment Quality. SAMtools has been used very heavily by a number of large human genome

TABLE 1. Description of fields in the SAM/BAM file format

<table>
<tr><th>Col</th><th>Field</th><th>Type</th><th>Regexp/Range</th><th>Brief description</th></tr>
<tr><td>1</td><td>QNAME</td><td>String</td><td>[!−? A−~] { 1,255}</td><td>Query template name</td></tr>
<tr><td>2</td><td>FLAG</td><td>Int</td><td>[0,216−1]</td><td>Bitwise flag</td></tr>
<tr><td>3</td><td>RNAME</td><td>String</td><td>* |[!−() +−<>−~] [!−~] *</td><td>Reference sequence name</td></tr>
<tr><td>4</td><td>POS</td><td>Int</td><td>[0,229−1]</td><td>1-based leftmost mapping position</td></tr>
<tr><td>5</td><td>MAPQ</td><td>Int</td><td>[0,28−1]</td><td>Mapping quality</td></tr>
<tr><td>6</td><td>CIGAR</td><td>String</td><td>* | ([0−9] +[MIDNSHPX=]) +</td><td>CIGAR string</td></tr>
<tr><td>7</td><td>RNEXT</td><td>String</td><td>* |=|[!−() +−<>−~] [!−~] *</td><td>Reference name of the mate/next segment</td></tr>
<tr><td>8</td><td>PNEXT</td><td>Int</td><td>[0,229−1]</td><td>Position of the mate/next segment</td></tr>
<tr><td>9</td><td>TLEN</td><td>Int</td><td>[−229 + 1,229−1]</td><td>Observed template length</td></tr>
<tr><td>10</td><td>SEQ</td><td>String</td><td>* |[A−Za−z=.] +</td><td>Segment sequence</td></tr>
<tr><td>11</td><td>QUAL</td><td>String</td><td>[!−~] +</td><td>ASCII of Phred-scaled base quality + 33</td></tr>
</table>

TABLE 2. Description of fields in the VCF file format

Col	Field	Description
1	CHROM	Chromosome name
2	POS	1-based position. For an indel, this is the position preceding the indel.
3	ID	Variant identifier. Usually the dbSNP rsID.
4	REF	Reference sequence at POS involved in the variant. For a SNP, it is a single base.
5	ALT	Comma-delimited list of alternative sequence(s)
6	QUAL	Phred-scaled probability of all samples being homozygous reference
7	FILTER	Semicolon-delimited list of filters that the variant fails to pass
8	INFO	Semicolon-delimited list of variant information
9	FORMAT	Colon-delimited list of the format of individual genotypes in the following fields
10+	Sample(s)	Individual genotype information defined by FORMAT

resequencing projects such as the Cancer Genome Atlas. Many additional features have been added to support simultaneous calling of variants in multiple samples including population allele frequencies, differential allele frequencies in two different groups of samples (association test), parent–child trios, and so on.

There is a good tutorial on the use of SAMtools for variant calling at http://samtools.sourceforge.net/mpileup.shtml.

The example uses the following SAMtools commands:

```
samtools mpileup -uf ref.fa aln1.bam | bcftools view -bvcg -> var.raw.bcf
bcftools view var.raw.bcf | vcfutils.pl varFilter -D100 > var.flt.vcf
```

By default, SNPs are called with a Bayesian model identical to the one used in MAQ. A simplified SOAPsnp model is implemented, too. Indels are called with a simple Bayesian model. The caller does local realignment to recover indels that occur at the end of a read but appear to be contiguous mismatches.

The `varFilter` filters SNPs/indels in the following order:

`d`: low depth
`D`: high depth
`W`: too many SNPs in a window
`G`: close to a high-quality indel
`Q`: low mapping quality
`g`: close to another indel with more evidence (indel only)

GATK

The Genome Analysis Toolkit (GATK) is a structured software library produced by the Broad Institute (McKenna et al. 2010) that contains tools for many types of analysis of NGS data, with a particular emphasis on SNV detection for human medical resequencing. GATK provides an extension of the SNV tools in SAMtools, implementing a pipeline that includes: mapping ⇒ quality score recalibration ⇒ multiple sequence realignment ⇒ SNP/index calling (DePristo et al. 2011). The Map/Reduce approach to data processing allows for efficient access to all reads that cover a specific genomic position (from a sorted, indexed BAM file) so that quality score recalibration and multiple sequence realignment can be performed rapidly at the locations of putative SNVs.

A tutorial for the use of GATK for variant detection is available, including sample data sets, at http://www.broadinstitute.org/gsa/wiki/index.php/Best_Practice_Variant_Detection_with_the_GATK_v3.

The tutorial assumes that raw sequence reads (in FASTQ format) have been aligned to a reference genome using the BWA aligner to produce alignments in sorted indexed BAM format. The recommended workflow first removes duplicates (reads or read-pairs with identical start and stop positions), realigns around indel sites, then recalibrates quality scores for each base in each read. Better results can be achieved by processing a large number of similar samples together, so that variants in different samples can contribute additional information, but this is computationally extremely demanding.

The reprocessed alignments are then used for variant discovery. A quality score confidence threshold of Q30 is used as a default, but this assumes >10× coverage of the genome. Variant calls are possible at lower coverage with a confidence threshold as low as Q4. The goal of the Basic SNP filters is to remove alignment artifacts from the data. The VariantFiltrationWalker flags SNPs within clusters (three SNPs with 10 bp of each other) and those in poorly mapped regions with more than 10% of the reads having mapping quality 0. The Variant quality score recalibration tool dynamically builds an adaptive error model using known variant sites, then applies this model to estimate the probability that each variant is a true genetic variant or a machine artifact (i.e., a false positive). This is a knowledge-based approach using, as training sets, priors for sequence variants derived from previous sequencing projects such as the HapMap and 1000 Genomes Projects.

It is worth re-emphasizing this point. SNP calling is very difficult. It is not possible to make highly accurate ab initio variant detection with acceptably low levels of false positives and false negatives. The profiles of true variants and common false positives must first be learned from a large set of similar data where the truth is known.

Cancer SNVfinder

Cancer SNVfinder is designed to infer SNVs from NGS data from tumors to address this problem. First, it models allelic counts as observations and infers SNVs and model parameters using an expectation maximization (EM) algorithm and is therefore capable of adjusting to deviation of allelic frequencies inherent in genomically unstable tumor genomes. Second, it models nucleotide and mapping qualities of the reads by probabilistically weighting the contribution of a read/nucleotide to the inference of a SNV based on the confidence of the base call and the read alignment. Finally, it combines filtering out low-quality data in addition to probabilistic weighting of the qualities.

REFERENCES

Altshuler D, Durbin RM, Abecasis GR, Bentley DR, Chakravarti A, Clark AG, Collins FS, De La Vega FM, Donnelly P, Egholm M, et al. (1000 Genomes Project Consortium). 2010. A map of human genome variation from population-scale sequencing. *Nature* **467:** 1061–1073.

Bravo HC, Irizarry RA. 2010. Model-based quality assessment and base-calling for second-generation sequencing data. *Biometrics* **66:** 665–674.

DePristo M, Banks E, Poplin R, Garimella K, Maguire J, Hartl C, Philippakis A, del Angel G, Rivas MA, Hanna M, et al. 2011. A framework for variation discovery and genotyping using next-generation DNA sequencing data. *Nat Genet* **43:** 491–498.

Li H, Durbin R. 2009. Fast and accurate short read alignment with Burrows–Wheeler transformation. *Bioinformatics* **25:** 1754–1760.

Li H, Ruan J, Durbin R. 2008. Mapping short DNA sequencing reads and calling variants using mapping quality scores. *Genome Res* **18:** 1851–1858.

Li R, Li Y, Fang X, Yang H, Wang J, Kristiansen K, Wang J. 2009. SNP detection for massively parallel whole-genome resequencing. *Genome Res* **19:** 1124–1132.

McKenna A, Hanna M, Banks E, Sivachenko A, Cibulskis K, Kernytsky A, Garimella K, Altshuler D, Gabriel S, Daly M, DePristo MA. 2010. The Genome Analysis Toolkit: A MapReduce framework for analyzing next-generation DNA sequencing data. *Genome Res* **20:** 1297–1303.

Sherry ST, Ward M, Sirotkin K. 1999. dbSNP-database for single nucleotide polymorphisms and other classes of minor genetic variation. *Genome Res* **9:** 677–679.

Sherry ST, Ward MH, Kholodov M, Baker J, Phan L, Smigielski EM, Sirotkin K. 2001. dbSNP: The NCBI database of genetic variation. *Nucleic Acids Res* **29:** 308–311.

WWW RESOURCES

http://samtools.sourceforge.net/mpileup.shtml SAMtools webpage.

http://sourceforge.net/projects/maq/files MAQ is a set of programs that map and assemble.

http://vcftools.sourceforge.net/specs.html VCF (Variant Cell Format) home page.

http://www.broadinstitute.org/gsa/wiki/index.php/Best_Practice_Variant_Detection_with_the_GATK_v3 GATK v3 tutorial.

9

ChIP-seq

Zuojian Tang, Christina Schweikert, D. Frank Hsu, and Stuart M. Brown

Chromatin immunoprecipitation (ChIP) is a method to study genome-wide protein–DNA interactions. ChIP can be used to determine the locations of binding sites on the genome of transcription factors and modified **histones**. One of its significant strengths is that it can determine locations on the genome for protein–DNA interactions in living cells or tissues, allowing an investigator to obtain a snapshot of these interactions at an experimentally important moment in time.

ChIP works by immunoprecipitating fragments of DNA using an antibody directed at a protein that has been cross-linked to the DNA. Historically, there are two types of ChIP, chromatin sheared by sonication, called cross-linked ChIP (XChIP), and native ChIP (NChIP), which uses native chromatin sheared by micrococcal nuclease digestion. Mapping transcription-factor-binding sites or weakly binding chromatin-associated proteins is usually performed with XChIP. NChIP is better for histone modifications.

The principle of this assay is that DNA-binding protein in living cells can be cross-linked to DNA with a reagent such as formaldehyde. The cells are then lysed, and DNA is sheared into small fragments (200–1000 bp) by either sonication or micrococcal nuclease digestion. At this point, immunoprecipitation is performed by using an antibody specific for a known DNA-interacting protein, to isolate **DNA fragments** bound to the target protein (Solomon et al. 1988). DNA fragments are then released by reversing the cross-links and purified. The purified DNA fragments are supposed to be associated with the protein of interest in vivo.

Conventionally, the DNA fragments obtained from ChIP experiments are assessed by PCR using gene-specific primers. The enrichment of DNA fragments in the region of interest is compared with PCR products from control samples, such as untreated genomic DNA or precipitation with an anti-IgG antibody. The advantage of this method is the use of quantitative real-time PCR to assess accurately the enrichment levels of interrogated sites. However, it is labor-intensive, and it does not allow screening for novel targets.

The introduction of **microarrays** allowed researchers to identify the DNA fragments targeted by ChIP by hybridization to a microarray, a method known as ChIP-on-chip (Huebert et al. 2006) or ChIP-chip (Kim et al. 2007). Because microarrays contain large number of probes of known genomic sequences, they allow a genome-wide view of DNA–protein interactions.

In place of a microarray, **next-generation sequencing** (NGS) technology, such as the Illumina Genome Analyzer, can be used to directly identify fragments enriched by ChIP. This method, known as **ChIP-seq**, was recently developed and is now widely used for identifying protein-binding sites on chromatin (Johnson et al. 2007; Robertson et al. 2007). ChIP-seq requires only 30- to 50-bp-long **sequence reads** to map fragments to the **reference genome** and identify binding sites. The first studies using ChIP-seq included localization of histone modifications in human T-cells (Barski et al. 2007) and in mouse embryonic stem (ES) cells (Mikkelsen et al. 2007) and localization of transcription factors in human T-cells (Johnson et al. 2007) and in human HeLa S3 cells (Robertson et al. 2007). Most DNA-binding proteins investigated are transcription factors, RNA polymerase–binding sites, and markers of chromatin structure (histone modifications, DNase hypersensitive sites). It is also possible to use ChIP-seq to directly study modifications of DNA molecules, such as patterns of DNA methylation.

Compared with ChIP-chip, ChIP-seq provides many advantages (Hoffman and Jones 2009; Park 2009; Liu et al. 2010). First, it provides higher resolution, so that the actual binding site of a factor can be identified within 10–30 bp on the genome (Kharchenko et al. 2008; Zhang et al. 2008; Hoffman and Jones 2009). Second, ChIP-seq does not require prior knowledge of potential binding sites—it allows for genome-wide novel binding site discovery. Third, for ChIP-chip, there are many uncertainties in the hybridization process, such as cross-hybridization between probes, which is not the case in ChIP-seq. Fourth, ChIP-seq allows binding site identification in the repetitive regions, which generally cannot be interrogated by ChIP-chip. There are ~10%–30% of binding sites within repetitive regions (Bourque et al. 2008). With 36- to 72-bp NGS sequence reads, there is ~80%–90% genome mappability. Reads can be mapped to sequences at the boundary region between unique sequence and repetitive sequence (Park 2009; Rozowsky et al. 2009). Five, ChIP-seq is cost effective with very deep sequencing. Chip-chip costs about $400–$800 per array for 1–6 million probes (Park 2009). Sometimes, multiple arrays will be needed to investigate a whole genome (Barrera et al. 2008). The cost of ChIP-seq is rapidly declining with improved sequence yield of NGS machines, so that multiplexing samples in a single lane is now possible (see Chapter 1). Sixth, the amount of input material required by ChIP-seq is much less than ChIP-chip. Normally, ChIP-chip needs 4–5 mg of materials; instead, ChIP-seq needs as little as 10 ng (Hoffman and Jones 2009). Given these many advantages, it is not surprising that ChIP-seq has become more and more widely used and has essentially replaced ChIP-chip as the standard technology for the study of DNA–protein interactions.

ChIP-seq can make use of any NGS technology, such as ABI-SOLiD, Roche 454, and Illumina Genome Analyzer, but Illumina has become the most popular sequencer in this field. For example, the Illumina HiSeq2000 sequencer generates up to 600 Gb per run. This means it can generate up to 230 million reads per lane at a cost of approximately $1000 per lane. It also supports multiplexing sequencing, which means that multiple samples can be sequenced on the same lane, reducing costs. This large data output requires significant computational resource and data storage capability (see Chapter 12).

ChIP-seq PEAK DETECTION

A ChIP-seq experiment detects short sequence tags derived from the ends of DNA fragments that have been cross-linked to a protein, which is the target of immunoprecipitation. The location of each tag is obtained from **alignment** with the reference genome. Because NGS reads of ~30 bases are adequate to uniquely determine the location of 70%–80% of tags on the genome, there is little value to using longer reads or paired ends. The data from the experiment, after filtering to remove data points that fail quality checks, consist of a list of the binding locations of each of the tags.

Analysis of these data is generally performed in one of two ways. The first makes use of the fact that the tag alignment locations specify the start of a run of bases (often 36 bases, but some machines produce longer runs). If tag starts are close together, then the tags will overlap and "pile up" over regions of the genome. A curve can therefore be created where the height at each location on the genome is equal to the number of overlapping tag starts. Because duplicate tags that start at identical genomic locations are filtered out to reduce the effects of PCR artifacts, the maximum possible value on the curve will be equal to twice the length of a tag (allowing for reads on both strands of the DNA).

The second approach is to assume that there is a function, say $P(x)$, that is the probability that a tag will bind starting at location "x" on the plus strand (or the minus strand) of a chromosome. In this model, the data collected samples this probability function, which can then be reconstructed yielding peaks in the probability curve. The larger the number of measurements, the better is the estimate of P. (This is a standard problem in several disciplines, most notably high-energy physics, where you have to estimate the cross section of a particle interaction given the relatively small number of events that your experiment measures.) Once you know, or have a sufficiently accurate estimate of $P(x)$, you use it to infer the locations of genes.

There are several ways to get a reasonable estimate for $P(x)$ given the limited number, N, of locations measured. One is to obtain a band-limited Fourier transform, which has the effect of smoothing the measurements to obtain an estimate of *P*. Another is to convolve the location measurements with the anticipated shape of the peak. This is computationally very fast and yields a curve that should show the location of all of the uncluttered peaks.

In both approaches, it is important to note that short tags come from the two ends of a longer segment of DNA (length *d*) protected by the binding protein. In consequence, in the region of a binding site, peaks will be seen for the "+" and "–" strand tags that are offset by approximately the length of the original protected strand length. The best estimate for the location of the target site is expected to be roughly in the middle between the "+" and "–" strand peaks. Shifting the "+" strand peak by $d/2$ to the right and the "–" strand peak by $d/2$ to the left and averaging them provides an improved estimate of the binding site on the genome.

The first step in a peak search task is to estimate the length *d* of the protected DNA fragment. For ChIP-seq experiments in which a transcription factor is the target of the immunoprecipitation, a straightforward way to do this is to use the transcription start sites (TSSs) of all of the genes in the genome (about 24,000 in the human) as the most likely binding locations of the protecting protein, then to create a histogram of binding locations for all of the "+" and "–" strand tags in a window centered at each of the TSSs. A typical result is shown in Figure 1. In this case, the best initial estimate of the separation is 136 bp, and this value can be used to compute peak locations, which can then be used in place of the TSS list to re-estimate *d*.

Once a curve is obtained (from pile-up, probability or convolution), peaks are immediately obvious. Two main questions remain: Which peaks are significant, and, for the peaks judged significant, which peaks most accurately indicate the locations of the genes. Evaluation of significance has been performed in two main ways: by testing whether the peak being evaluated is statistically significant based on the number of tags bound in the local region, or whether it is significant based on the number of tags found from a control experiment's data set designed closely to mimic the experiment. These methods can be combined if the investigator is particularly concerned about an adverse false discovery rate. Some peak-calling packages have the evaluation of the statistical significance of the peaks built-in and therefore require a control data set to be available for them to function, whereas others do not. One benefit of keeping the peak identification step and the evaluation of significance separate is that the peaks can be evaluated and winnowed down against multiple control data sets.

Peak-Calling Packages

There are numerous peak-finding software packages available, and they have been reviewed in depth in several papers. The use of different software to analyze a single data set can produce strikingly different results. As noted in these publications, the problem with method evaluation is that there are no real-world data sets that provide a gold standard for peak calling with which these methods can be compared. In

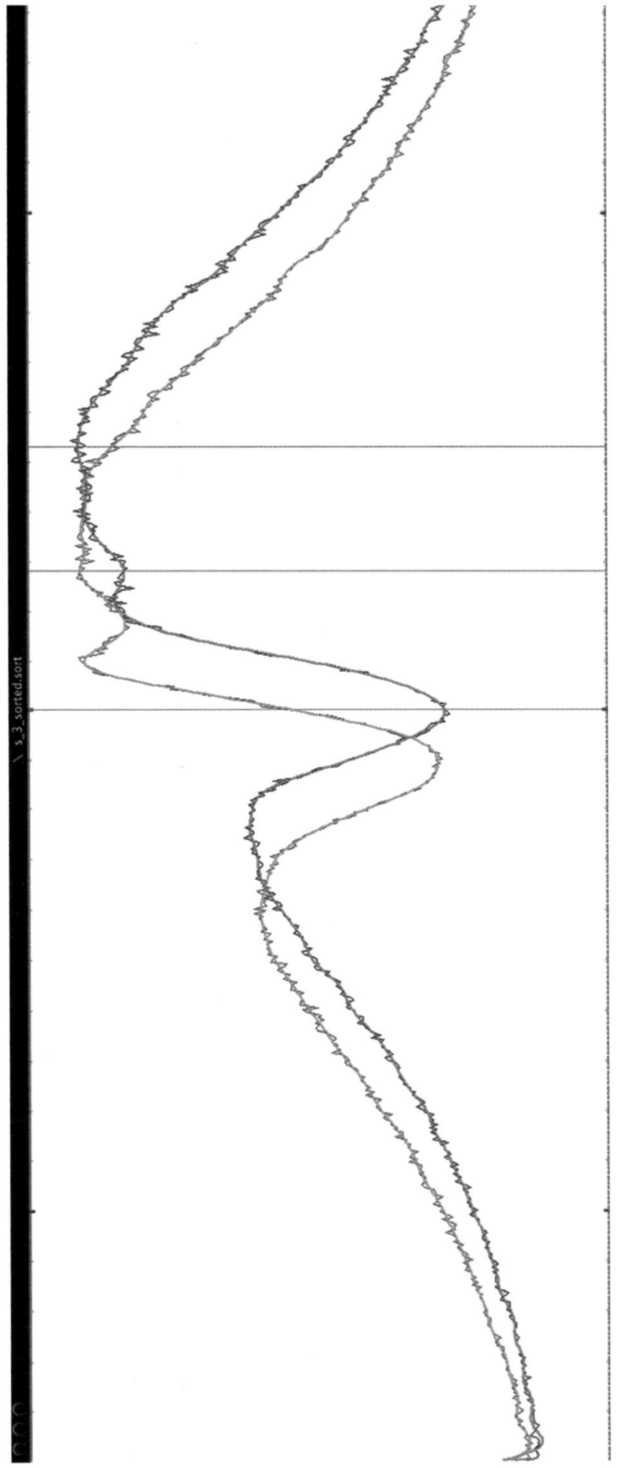

FIGURE 1. A graph of the mapping position of + (blue) and − (green) strand reads with respect to TSS position across all genes in the genome. These data are from a human H3K4 ChIP-seq sample and show a peak of reads at the TSS, a separation of + and − strand reads, and a clear dip in reads at nucleosome-free sites ~150 bp upstream of the TSS. (Figure created by Phillip Ross Smith.)

principle, synthetic data sets should provide a means to test and compare peak-calling **algorithms**, but in practice, it is found that these data sets do not provide much discriminatory power. Wilbanks and Facciotti's study (Wilbanks and Facciotti 2010) tested 11 packages against previously published ChIP-seq data for three experiments (NRSF, GABP, FoxA1) and showed that with default settings for the tuning parameters of each of the packages, the number of peaks called varied by a factor of 3×–4×. The set of peaks called in common across all 11 packages amounted to only about one-tenth of the peaks found by some packages.

Since it was introduced in 2008, the MACS algorithm (Zhang et al. 2008) has achieved broad usage and has been seen to perform well in side-by-side comparison tests. However, the clear message from the comparisons is that the choice of peak-calling algorithm needs to be one that is made by the investigator on a case-by-case basis and that there continues to be room for work to seek algorithmic improvements. In the latter part of this chapter, we work through a data analysis using MACS, but in our own work, we have successfully developed our own peak-calling software with good effect.

As in all NGS experiments, the amount of data that must be processed and interpreted is dauntingly large. Given a list of possibly tens of thousands of potential peaks, the investigator needs additional criteria to evaluate the results. These criteria must generally be drawn from the biological problem that motivated the experiment. In addition, it should be noted that any algorithm for peak calling involves some type of thresholds, be they set as parameters or computed from the data. Inevitably, some peaks receive scores just above or below these thresholds. When making a biological comparison between two samples, treatments, conditions, etc., these near-threshold peaks will be a source of variability and, if care is not taken, will be added to the list of differentially regulated genes. ChIP-seq, like any bioassay, requires replicates and careful statistical analysis.

Algorithm Implementation

In our laboratory, we have explored two new approaches to the peak-calling problem. In the first, we have built a computationally fast package to detect and visualize peaks based on the probabilistic model for tag binding. Initial convolution of the data with a convolution kernel, such as the one used in the MACS program (but that can be designed by the investigator using criteria appropriate for the experiment), yields a score at each tag-binding location. Peaks in the scores provide estimates for tag peak locations. Fourier smoothing of the tag-binding locations in the neighborhood of the score peaks allows more reliable estimates to be made of P and the peak locations. Each of the locations of a peak in P can then be tested for statistical significance using one or more criteria, including a comparison with suitable background control data sets.

Processing takes place in three steps; first the scoring of the data set; second, the choice of an initial peak height cutoff using a visualization tool that displays data around a peak and that can be used to drill down at areas of interest; and finally, a peak-calling program that uses the initial peak location estimates to refine the peak locations and assess their statistical significance.

Basic analysis of ChIP-seq data is relatively straightforward after sequences are generated from NGS sequencers. It includes QC, mapping sequence reads to a reference genome, and peak finding. Sequences are mapped to a known reference genome using a stringent **short read alignment algorithm**. Illumina provides its own alignment software called CASAVA (Consensus Assessment of Sequence and Variation). CASAVA version 1.8, released in June 2011, can align sequences in lengths of 30–200 bp in parallel with error rates < 1%–2%. There are also many publicly available alignment programs such as SOAP (Short Oligonucleotide Alignment Program) (Li et al. 2008a), SeqMap (Jiang and Wong 2008), MAQ (Mapping and Assemblies with Qualities) (Li et al. 2008b), RMAP (Smith et al. 2008, 2009), ZOOM (Lin et al. 2008), Novoalign (http://novocraft.com) for not-for-profit projects within not-for-profit organizations, Bowtie (Langmead et al. 2009), BWA (Li and Durbin 2009), SOAP2 (Li et al. 2009b), and many others (see Chapter 4). The input file for most of the software is **FASTQ format**, which is standard output from Illumina's pipeline. Alignment parameters vary among different software, but most methods allow one or two mismatches per seed (16–32 bp) caused by sequencing error and SNPs/**indels**. Sequence reads that do not map to a unique position on the genome are generally discarded. More sensitive (less stringent) DNA alignment methods generally do not improve the quality of results from a ChIP-seq experiment. Most short read alignment software saves results in SAM or **BAM file** format (Li et al. 2009a). This type of file can be viewed in most visualization software, such as GenomeStudio Software (Illumina Inc.), GBrowse (Stein et al. 2002), IGV (Robinson et al. 2011), IGB (Nicol et al. 2009), GenomeView (http://genomeview.org/), and many others.

There are several sources of bias in ChIP-seq data (Barski and Zhao 2009; Liu et al. 2010).

1. *Bias in the library construction.* Highly transcribed, or histone-free regions of chromatin may be preferentially fragmented by sonication or DNase. A preference for sequencing GC-rich regions in fragment selection has been observed (Dohm et al. 2008; Quail et al. 2008).
2. *Bias in the PCR step.* Some **sequence fragments** may be preferentially amplified, creating a huge number of replicated reads. To solve this bias, most analysis pipelines will remove duplicated reads.
3. *Bias in nonspecific regions.* Nonspecific interactions of the ChIP target protein with DNA, coprecipitation of unbound DNA fragments, or various types of contamination and amplification artifacts may occur. This can cause some regions

to be enriched in ChIP-seq data in a manner unrelated to the binding of the target protein. This bias can be removed by using a control library, such as a mock ChIP using IgG antibody or untreated "input DNA." The use of computational background models such as the binomial model or **Poisson model** are much less effective in removing nonspecific bias.

4. *Bias around TSS regions.* For ChIP-seq applications, even the control library tends to have more enrichment around TSS regions because of local chromatin structure, DNA amplification, or genome copy number variation (Zhang et al. 2008; Vega et al. 2009). This problem can also be reduced by using a control library.

Once the reads are mapped to the genome, researchers are interested in finding the genomic regions that indicate protein-binding (or histone-modification) positions enriched by significant numbers of reads. There are both commercial and freely available peak-finding software for finding these regions. Simple algorithms can identify peaks as windows of the genome sequence that contain counts of mapped tags greater than a background threshold (Johnson et al. 2007; Robertson et al. 2007). Peak-finding software finds significantly enriched regions using either a single ChIP-seq data file with randomly distributed background models (such as Poisson or binomial models) or paired files of ChIP-seq data with control background data. There are several different methods to use a control sample. It is possible to normalize read counts per genomic interval (windows) and then directly subtract control reads from ChIP sample reads, calculate ratios between sample and control within each window, or find peaks in the control in order to set a false discovery threshold for the sample. Rozowsky et al. (2009) found that control samples contained clusters of aligned sequence reads at similar locations to peaks in ChIP samples, possibly caused by relaxed histone structure at TSS, creating enhanced susceptibility to DNase digestion and physical shearing at these sites. Nix et al. (2008) have made detailed observations of background peak distributions and false discovery rates that confirm the nonrandom distribution of sequence tags in ChIP-seq control lanes. Without background correction with a control sample, ChIP-seq peak detection methods will have a high rate of false-positive results.

Because the genomes of complex eukaryotes are billions of bases long, accurately and reproducibly identifying and quantifying the heights and/or areas of ChIP-seq peaks require a sophisticated computational solution. Many computational methods have recently been developed for the analysis of ChIP-seq data including GenomeStudio (Illumina Inc.), FindPeaks (Fejes et al. 2008), peak detection based on overlaps of extended sequence reads (Robertson et al. 2007), ChIP-seq Peak Finder/E-RANGE (Johnson et al. 2007; Mortazavi et al. 2008), SISSRs (Jothi et al. 2008), CisGenome (Ji et al. 2008), QuEST (Valouev et al. 2008), USeq (Nix et al. 2008), and MACS (Zhang et al. 2008). Several reviews have shown list of software available

(Barski and Zhao 2009; Hoffman and Jones 2009; Park 2009; Pepke et al. 2009; Liu et al. 2010). Wilbanks and Facciotti (2010) and Laajala et al. (2009) have provided good evaluation of peak-finding software. Overall, different software packages detected dramatically different numbers of peaks using default parameters. When parameters were adjusted to produce similar numbers of detected peaks, overlap of peak locations detected by the different methods was poor. The continued proliferation of ChIP-seq peak-finding software is an indication that no well-validated general method exists that performs well in the hands of most investigators.

The comparison and benchmarking of ChIP-seq analysis software is itself a difficult bioinformatics problem. One approach is the correlation of reported peaks with external data sources such as RefSeq gene annotations, under the assumption that valid ChIP-seq peaks (for most transcription factors and many histone modifications) will be associated with the promoters of known genes (see Fig. 2). Although this validation method is far from perfect, each software package can have its parameters optimized to find the largest number of peaks near gene transcription start sites (TSSs) with the smallest fraction of peaks located far from TSSs (high sensitivity, low false discovery). In some cases, it is also possible to correlate ChIP-seq peaks with ChIP-chip data for the same experimental system. However, neither of these external data sets provides an adequate "gold standard" against which to measure the specificity and sensitivity of peak-detection results. Published results from ChIP-seq experiments show only 30%–40% correlation of peak detection with ChIP-chip binding site predictions (Robertson et al. 2007; Ji et al. 2008). We have studied ChIP-seq peak-detection software based on the reproducibility of peak detection across technical replicates. Even this approach does not provide a true measure of accuracy for peak-detection methods because we cannot empirically determine if a peak-detection disagreement between replicates or between different software is caused by a false-positive or a false-negative result. Ideally, all software should be validated against common "gold standard" data sets composed of a large number of replicates of ChIP-seq data from biologically well-studied systems in which all of the binding sites are known (Håndstad et al. 2011), and synthetic data sets in which peaks are inserted into control sequence.

MACS (Model-Based Analysis of ChIP-seq) (Zhang et al. 2008) first models the shift size of ChIP-seq tags empirically using bimodal enrichment pattern, with plus strand tags enriched upstream and minus strand tags enriched downstream, to better locate the precise binding sites. It then identifies the candidate peaks with a significant tag enrichment based on a Poisson distribution p-value. It accepts either single ChIP samples or paired sample and control data (such as IgG or input DNA). For the control samples, it uses a dynamic parameter, λ, for each peak region to capture the local bias in the genome. It outputs candidate peaks with p-values below a user-defined threshold with the ratio of fold enrichment between the ChIP-seq read count and local λ. For experiments with control data available, it also outputs

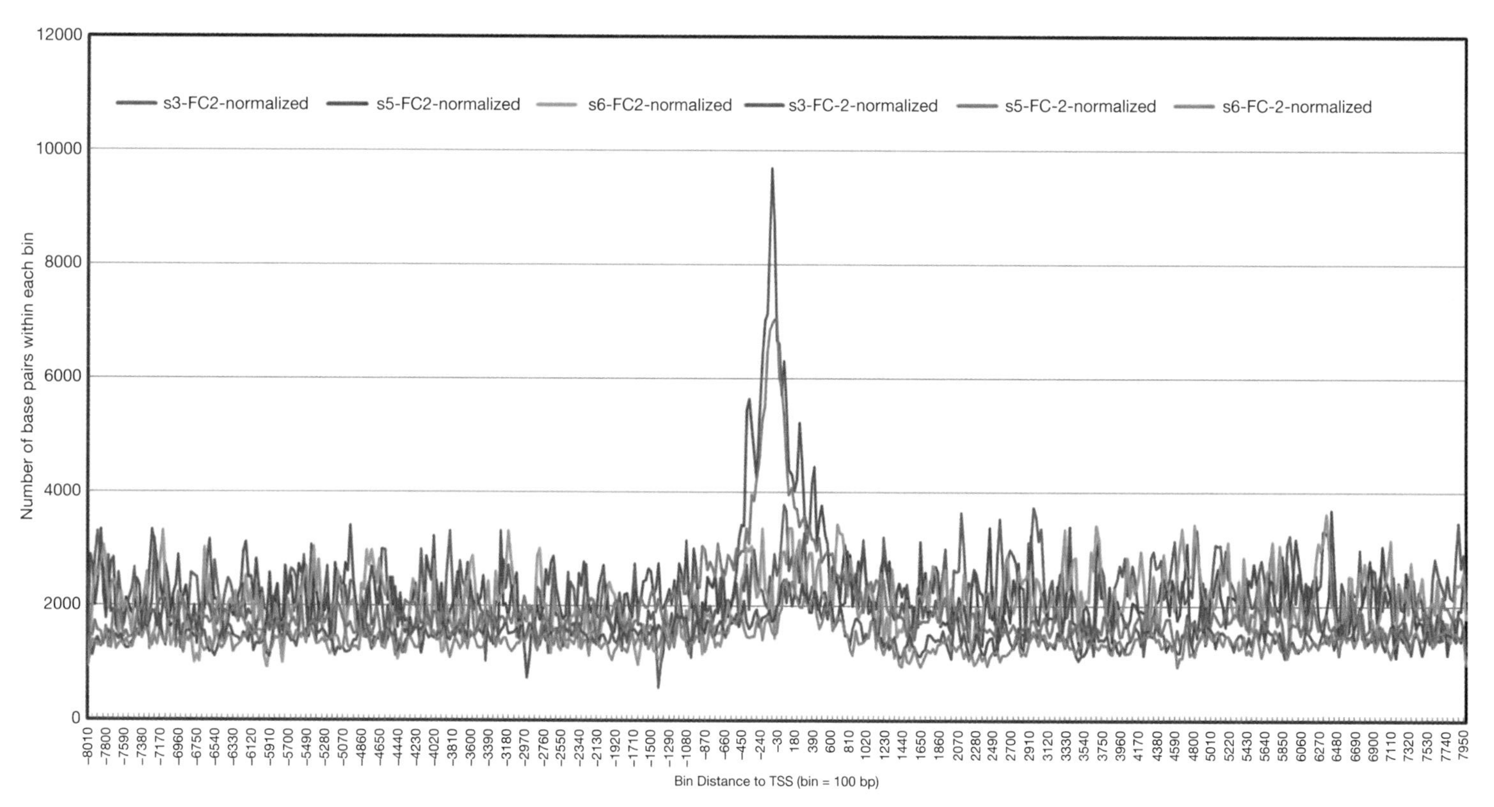

FIGURE 2. Two ChIP-seq samples (red and blue lines) show a clear clustering of reads near the TSSs of genes (in aggregate across the whole genome), with a maximum at approximately position −30, whereas reads from the other four samples do not have reads clustered near the gene TSSs.

FDR defined as (number of control peaks)/(number of ChIP peaks). MACS has become the most widely used peak-calling software, and it has been incorporated in the popular Galaxy web-based genomics toolkit as the standard method for ChIP-seq peak calling.

The commercial GenomeStudio application (Illumina, Inc.) uses a simple window/threshold method to detect peaks in the sequence reads mapped to a reference genome by the ELAND (Efficient Large-Scale Alignment of Nucleotide Databases) alignment method. In our evaluation, windows of 400 bp and tag thresholds of 5–10 tags/window were optimal to identify both transcription factor binding and histone modification sites. This method does not provide for any form of background (control sample) correction, nor does it provide a false discovery rate (FDR) or *p*-value calculation to support predicted peaks. Reproducibility of peak region detection across replicates varied from 38% to 90% depending on depth of sequencing (tag density) and the type of antibody target used for the ChIP (see Table 1).

FindPeaks (Fejes et al. 2008) is one of the first open source programs for Chip-seq peak finding. It directly counts overlapping sequence reads (reads may be extended to the length of the original DNA fragments isolated by ChIP) and then calculates peak height as the maximum number of overlapping tags at any one base position within an enriched region (similar to Robertson et al. 2007). Optionally, it requires that the peak height be greater than a false discovery threshold value set by a Monte Carlo simulation of sequence reads randomly distributed over the genome. We have observed that a Poisson or Monte Carlo model does not accurately represent the background distribution of sequence reads from input DNA or anti-IGG antibody. FindPeaks does not provide a method to subtract background or to compare experimental and control samples.

ChIP-seq Peak Finder (Johnson et al. 2007) clusters the reads and uses the ratio of the counts in the ChIP versus the control sample to call peaks.

PeakSeq (Rozowsky et al. 2009) uses input-DNA control data to refine the selection and scoring of peak regions in Chip-seq experiments to improve the identification of transcription-factor-binding sites. Because it has been observed that signal peaks in the control data are highly correlated with potential binding sites, PeakSeq compensates for this signal, caused by open chromatin structure, with a two-pass strategy. PeakSeq first identifies enriched peaks in the Chip-seq data as candidate

TABLE 1. Peak detection reproducibility

Experiment	Tag correlation	PeakFinder	SISSRS	GenomeStudio
E2F4	0.86	17.2%	23.4%	38.0%
Sin3b	0.76	20.7%	16.0%	56.1%
H3K4	0.98	66.4%	25.6%	90.0%

regions. These putative regions are then compared with the normalized control, and the regions that are significantly enriched with mapped sequence tags relative to the control are identified as binding sites. The PeakSeq software is available at http://info.gersteinlab.org/PeakSeq.

Site identification from short sequence reads (SISSRs) estimates high read counts using Poisson probabilities and calls regions where the peaks shift from the forward to the reverse strand (Jothi et al. 2008). The SISSRs method is attractive because it explicitly makes use of information from the orientation of tags around a protein-binding site, where it is expected that forward-strand tags will be found upstream of the true binding site and reverse-strand tags downstream. This allows for very precise prediction of the binding site location. However, this method does not perform well for regions of low-tag density or for histone methylation ChIP, where tags are not neatly oriented. SISSRs tends to create many small peaks across large enriched regions, which are not reproducible across replicates.

CisGenome (Ji et al. 2008) uses a two-pass algorithm for peak detection to ensure adjustment for DNA fragmentation length. It can analyze both ChIP-seq and ChIP-chip data, or combine the two. To correct many types of systemic bias created by sample preparation, amplification, sequencing (or hybridization), and alignment, it uses both a ChIP sample and a control sample (input DNA or mock-ChIP with IGG) to compute FDR at each specific location. It also provides methods to detect binding regions, peak localization, and filtering. CisGenome is Windows software with a well-designed GUI. It also includes a genome visualization tool and a sequence motif finder for detailed analysis of transcription binding sites, including the ability to generate sequence logos from a collection of sites identified as ChIP-seq peaks (http://www.biostat.jhsph.edu/~hji/cisgenome).

QuEST (Valouev et al. 2008) provides a data-driven statistical analysis model to generate peak calls by leveraging the key attributes of the sequenced and aligned DNA reads, such as directionality (strand orientation) and the original size of ChIP-isolated DNA fragments. The statistical framework used is the kernel density probability estimation approach, which facilitates the aggregation of signals originated from densely packed sequence reads at protein interaction sites. Statistical estimation of FDR in QuEST requires two samples of control data for each experimental sample, which has proven inconvenient for most investigators.

The USeq software package (Nix et al. 2008) uses control data to calculate window summary statistics with four different methods (window counts, window counts with subtracted control counts, normalized differences, p-value based on random binomial distribution). Prefiltering is applied to remove all regions of the genome with significantly elevated p-values in control data (background peaks). Despite careful attention to FDR, window thresholds may be obtained by the application of inappropriate p-values. True peaks located in the same genomic regions as background peaks will not be detected regardless of peak height.

The peak-detection abilities of existing ChIP-seq software packages vary greatly depending on the type of protein that is targeted by the antibody used in the ChIP. Transcription factors, such as E2F4, bind strongly to a specific DNA sequence (a motif) near the transcription start site of a gene, producing distinct ChIP-seq peaks ~400 bases wide (twice the size of ChIP DNA fragments), with oriented tags that approximately follow a normal distribution (Fig. 3A). A second pattern is observed with transcription factors, such as Sin3a, that bind weakly to DNA together with cofactors, yielding wider ChIP-seq peaks (800–1600 bases) with a flat distribution of lower tag density, and unoriented tags (Fig. 3B) and greater variability among replicates. Modified histone proteins, such as H3K4-3met, are a third type of ChIP-seq target that produce much wider peaks (~4000 bases) and unoriented tags (Fig. 3C).

The reproducibility of peak calls is particularly important to many investigators who wish to use the ChIP-seq technique to investigate biological differences—to discover how transcription factor binding or histone methylation changes in response to various developmental, environmental, genetic, or other factors. To detect meaningful biological differences, the peak-finding method must be both sensitive and

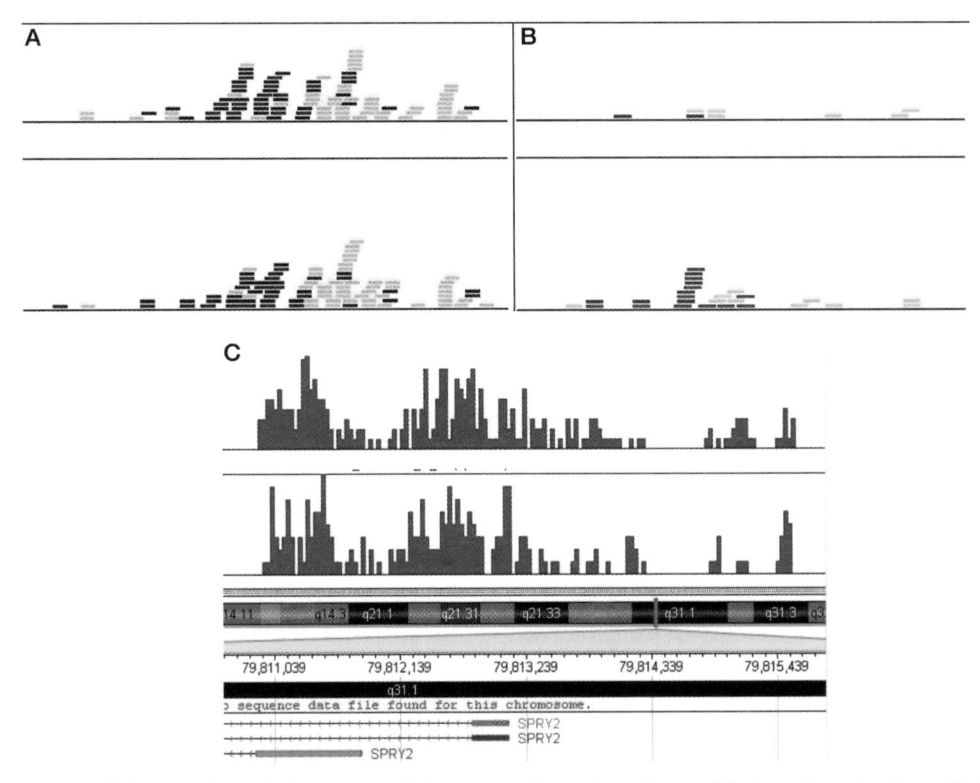

FIGURE 3. (*A*) A single peak from E2F4 ChIP-seq showing oriented tags. (*B*) A peak region from Sin3b ChIP-seq showing weaker more dispersed binding. (*C*) A peak region from H3K4-3met histone ChIP-seq showing a much larger span of enriched tags extending downstream beyond the first **exon** of gene *SPRY2*.

robust (able to produce consistent results across replicates). A peak-detection technique that only identifies the strongest signals from binding sites that are well supported by many overlapping sequence tags will not be adequate to dissect subtle genome-wide changes.

ILLUMINA ChIP-seq SEQUENCING

Illumina ChIP-seq data produced from the GAII, GAIIx, or HiSeq machines are processed in several phases to prepare them for ChIP-seq analysis.

1. *Image analysis and base calling.* The actual sequence reads are generated from the images acquired during **sequencing by synthesis** chemistry on the Illumina machine. Image analysis and base calling with Illumina Real Time Analysis software takes place while the sequencer is operating.
2. *The short sequence reads are aligned to the genome.* The ELAND algorithm is provided as part of the Illumina CASAVA pipeline (current version 1.8). ELAND outputs a standard alignment file, called sampleId_Index_Llane#_Read#_serial#_export.txt.gz, for each sample, which lists all of the sequence reads with their respective genomic coordinates. CASAVA can also output a sequence file in FASTQ format without alignment for each sample. This sequence FASTQ file is called sampleId_Index_Llane#_Read#_serial#.fastq.gz. All of the sequence reads are shown in FASTQ files.
3. *The CASAVA pipeline generates a results summary file, called the Flowcell_Summary_FCID.htm file.* This file contains comprehensive results and performance measures of your analysis run.

Given the output from Illumina above, the following additional quality analysis is recommended.

1. Genome-wide measures of overall clustering of sequence reads on the genome are independent of any software used for predicting peaks. A quick look at these values can predict if significant peaks (protein-binding sites) can be found in the data set, and therefore if time should be spent optimizing the parameters of the peak-finding software.
   a. Read distance distribution is a measurement of the overall clustering of reads. A data set with many strong and wide peaks will show a large number of overlapping reads, whereas a set of background reads should show a random spread of reads across the genome, with little overlapping and an even distribution of distances between adjacent reads. This information is best understood as a read distance distribution graph. The x-axis represents the distance between one read and its next adjacent read. If two reads are located at exactly the same location on the genome, the distance between these two reads will be 0. The y-axis

represents the percentage of the reads with the certain distance. Interestingly, all of these graphs show a peak at an inter-read distance of 1 bp and a gradual decline as distances increase. The skewness of this read-distribution graph toward short distances is a direct reflection of the quality and intensity of peaks on the genome that predict protein–DNA interaction sites. In cases in which the input DNA shows a skewness to short inter-read distances on this graph, then there are visible peaks along the genome, which one would not expect for input DNA. Using input DNA with this type of distribution as a background for peak prediction will tend to block the identification of real peaks in the ChIP experimental data. A graph of a theoretical Poisson distribution of reads would have no skewness to short overlap distances (i.e., no peaks), but in our experience, every real background DNA sample has some peaks and a slight rise in the read distribution graph at short distances (see Fig. 4).

b. Genome **coverage** distribution is a measurement of the reads with respect to the genome. Coverage can be visualized on a graph where the *x*-axis represents the depth of coverage of each nucleotide position on the genome and the *y*-axis represents the percentage of nucleotides in the genome with each level of coverage. A data set with many strong and high peaks will show a high percentage of coverage, whereas a set of background reads should show a random spread of reads across the genome with a lower percentage of deeply covered positions. Two data sets (ChIP and Control) with similar read distance distributions may have significant peaks identified if they have different genome coverage

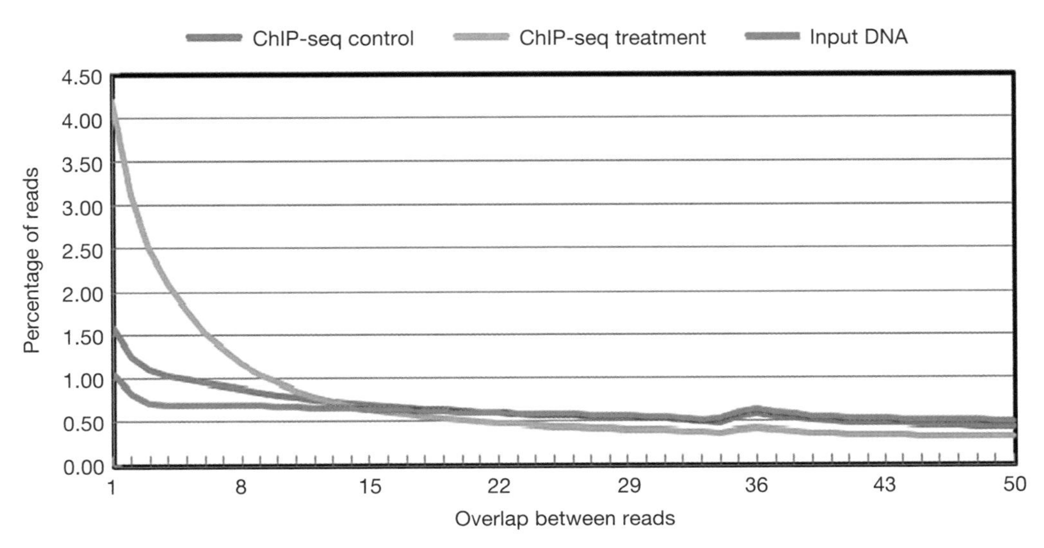

FIGURE 4. Read distance distribution graph for an experiment with strong ChIP peaks in a set of ChIP-seq treatment, control, and input DNA samples. A random (or unselected) distribution of reads across the genome will have little overlapping (assuming low genome coverage, such as 10–20 million 36-bp reads), whereas a strong ChIP signal will show a much greater percentage of overlapping reads.

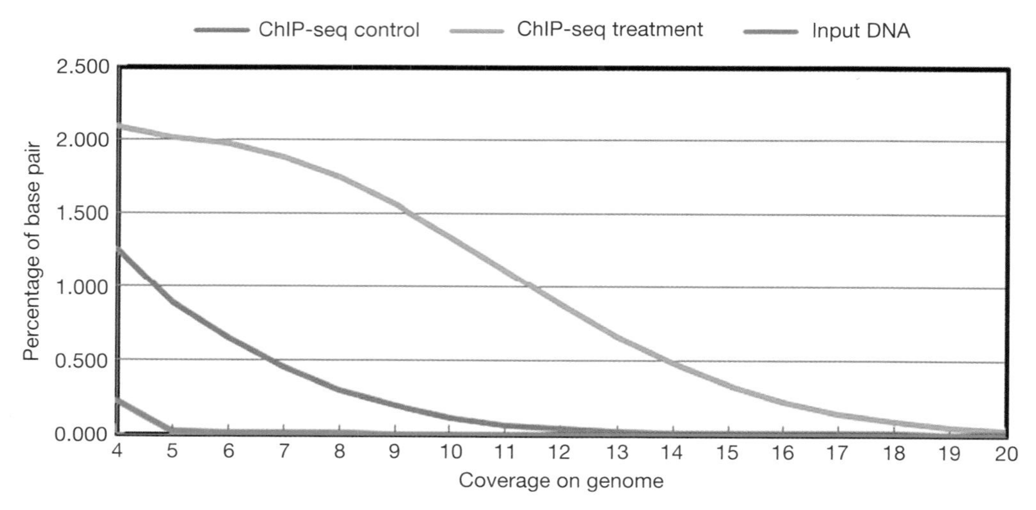

FIGURE 5. A genome coverage distribution graph of the same ChIP, control, and input data shown in Figure 4. The ChIP treatment sample shows deeper coverage for a higher percentage of the genome, whereas the input sample shows very few genome positions covered deeper than 4×.

distributions. A data set with significant peaks will have a coverage distribution shifted upward and to the left on this graph (see Fig. 5).

In some cases, the read distance distribution graph shows no clear difference between ChIP treatment and input DNA (Fig. 6A), but the genome coverage graph shows that there are still significant peaks to be found in the ChIP sample (Fig. 6B).

Another pattern is observed for ChIP experiments in which the targets affect a large portion of the genome, such as epigenetic studies that target modified histones. Large numbers of nonspecific peaks (wider and lower in height) are found for ChIP-seq data using input DNA as background (see Fig. 3C). This leads to a larger percentage of overlapping reads for the ChIP sample in the read distance distribution graph (Fig. 7A), but no clear difference in the genome coverage graph (Fig. 7B). In some experiments, ChIP fails to isolate unique DNA–protein interactions, leading to read distance and genome coverage graphs that show no difference between ChIP and input DNA samples (Fig. 8).

2. Statistical analysis for the percentage of read duplications.

Once QC steps are complete, the data can then be analyzed by peak-finding software and downstream annotation and analysis of peak locations on the genome.

1. *Peak finding.* Locally developed (such as TRLocator; https://sourceforge.net/projects/chipseqtools/files) and publicly available peak-finding software is used to find significant peaks. This is the most important and difficult step for ChIP-seq data analysis.

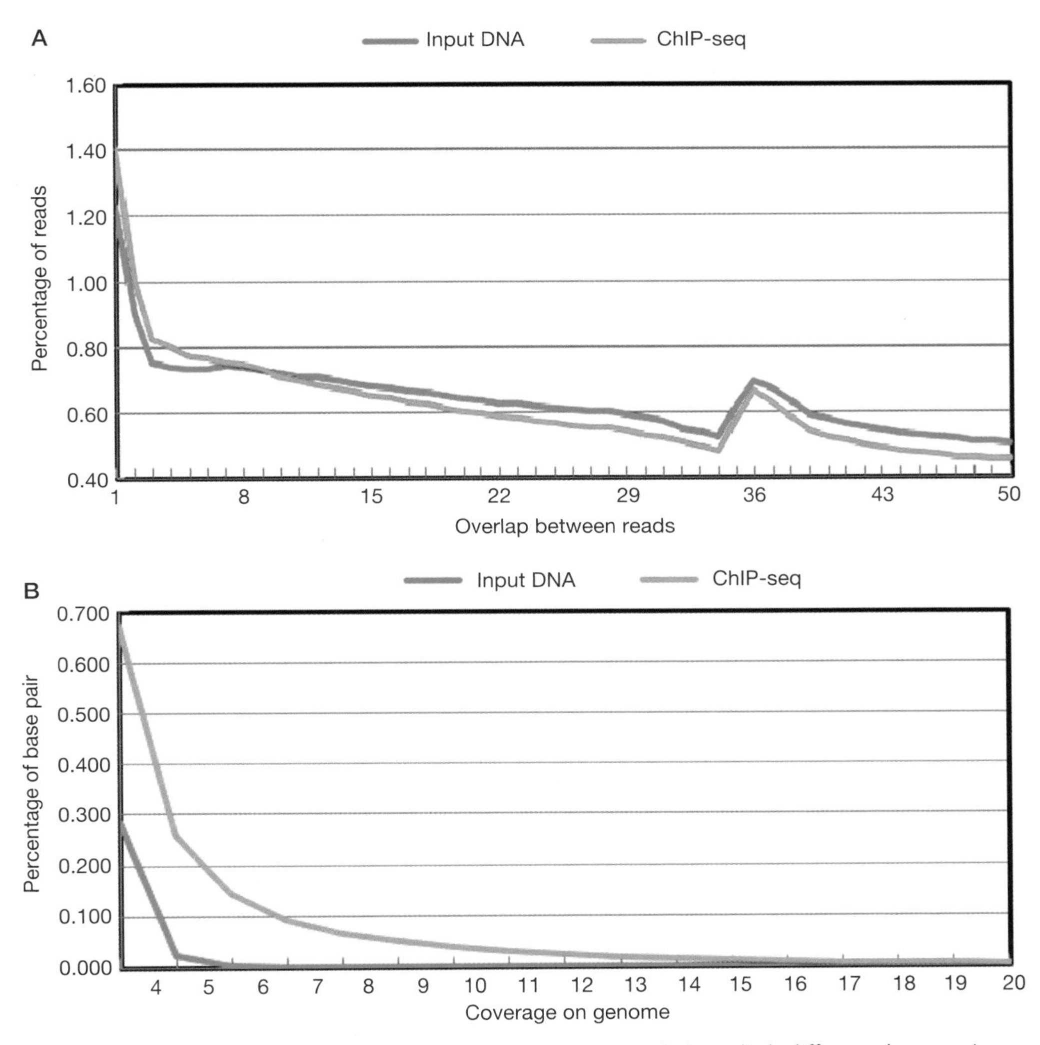

FIGURE 6. (*A*) In this experiment, the read distance distribution graph shows little difference between input DNA and ChIP sample. (*B*) However, the genome coverage graph shows that some genomic positions are more deeply covered by the ChIP sample, indicating that significant peaks can be found in the data.

2. *Peak annotation.* In-house developed software is used to annotate peaks.
3. *Binding site distribution.* This provides distribution of binding site distance to the TSSs.
4. *Comparison of binding sites.* This compares the difference of binding sites among the different biological treatments.
5. *Genome-wide/local region peak distribution.* This provides peak distribution genome-wide or locally given the gene expression data information.

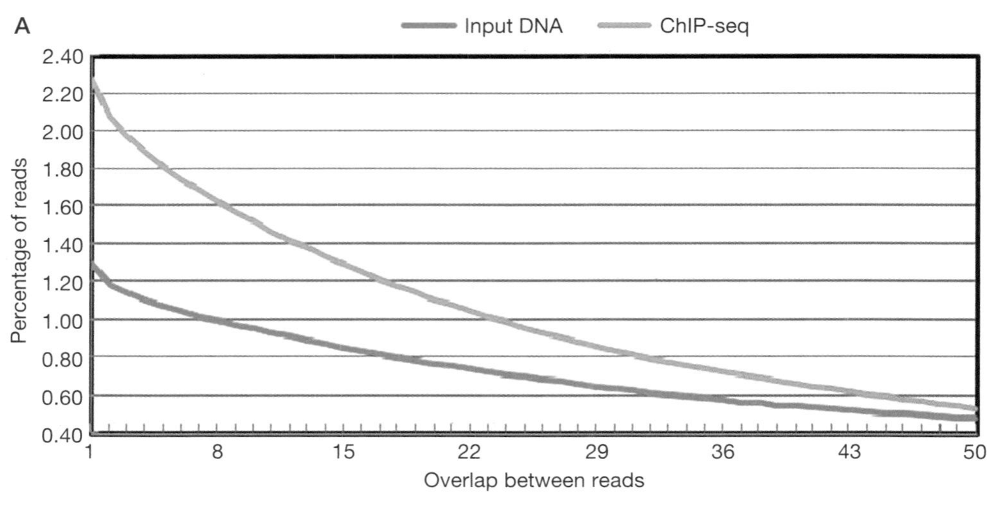

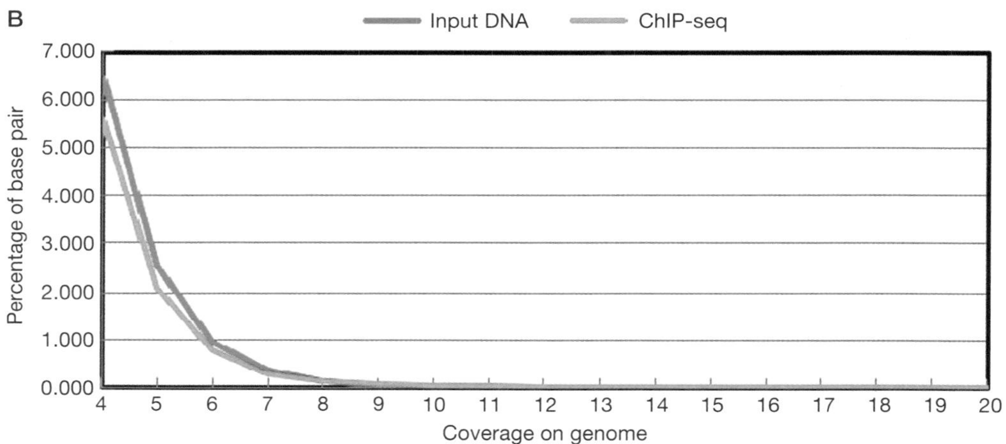

FIGURE 7. (*A*) ChIP-seq for methylated histone (H3K4) produces a pattern of high overlap between reads on the read distance graph compared with input DNA. (*B*) The genome coverage distribution graph shows little difference between ChIP and input samples.

TUTORIAL AND SAMPLE DATA

In this section, we provide a tutorial on how to analyze ChIP-seq data generated from Illumina's Genome Analyzer.

Normally, Illumina's Real Time Analysis (RTA) software will be used to do real-time base calling while the sequencing machine is running. If a Phix control lane is included, this will be used by HCS/SCS for Matrix and Phasing estimation. HCS/SCS will compute a median matrix only based on the tiles from this lane. Otherwise, HCS/SCS will compute these values using all of the tiles of the flow cell.

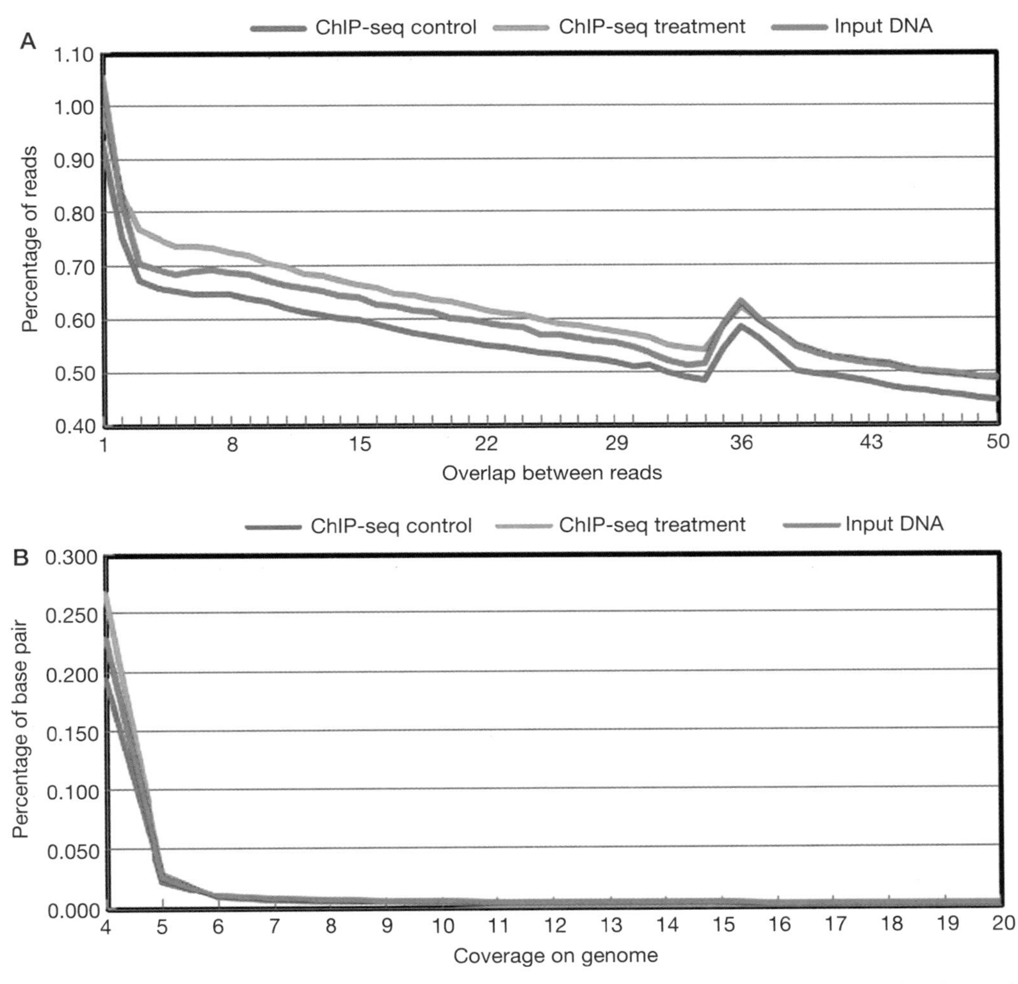

FIGURE 8. No difference is seen between ChIP-treatment, ChIP-control, and input DNA for the read distance distribution (*A*) or the genome coverage (*B*).

The primary output files of RTA from HiSeq or Genome Analyzer are binary base call files (*.bcl files) and *.filter files. Each *.bcl file contains base calls and quality scores per cycle. *.filter files specify whether a cluster passes the QC filters or not. RTA will also output *.stat files, which contain aggregated statistics for each cycle, and *.locs or *.clocs files, which contain the *x*,*y* position for every cluster. RTA is normally set up to transfer these files to a remote server during the sequencing run.

The first step of primary analysis is to convert bcl files into FASTQ files, which are the input files required by Illumina's downstream analysis software called CASAVA (current version 1.8).

The second step is to evaluate the quality of the sequencing run. The Illumina CASAVA program produces a file called Summary.htm file. This file includes the number of raw clusters generated, percentage of raw clusters that passed filtering, percentage of passed filtering clusters aligned to the reference genome, and the percentage of error rate (percentage of mismatches of aligned clusters) (Table 2).

Most researchers will analyze ChIP-seq experiment using the FASTQ-formatted output file. Most ChIP-seq experiments will use single-end sequencing. In this tutorial, we use two FASTQ files as initial files. One FASTQ file is from a ChIP sample, and the other FASTQ file is from the background control, which is input DNA in our case.

Alignment of FASTQ Files

FASTQ-formatted sequence files are common input files for most alignment software, such as BWA, Bowtie, and Illumina's ELAND. In this tutorial, we use FASTQ files that are generated by Illumina's CASAVA pipeline as input. We then align sequences to the reference genome using BWA as the alignment software.

Preparation of Reference Genome

Alignment requires a reference genome. Reference genomes can be downloaded in **FASTA format** from the NCBI website (ftp://ftp.ncbi.nih.gov/genomes) or the UCSC website (http://hgdownload.cse.ucsc.edu/downloads.html). BWA requires that the reference file be indexed to create a hash table of ***k*-mers**. The `bwa index` command will index the human genome hg18:

```
bwa index -a bwtsw Homo_sapiens_assembly18.fasta
```

BWA Alignment

The first step of BWA alignment is to find the suffix array (SA) coordinates of good hits of each individual read. The `bwa aln` command matches each read to the index:

```
bwa aln -m 10000000 -t 8 -I -f
s_7_sequence.txt.sai
Homo_sapiens_assembly18.fasta
s_7_sequence.txt
```

The next step of BWA alignment is to convert SA coordinates to chromosomal coordinates and generate alignments in the SAM format. The `bwa samse` command

TABLE 2. CASAVA summary file

Lane	Lane yield (kb)	Clusters (raw)	Clusters (PF)	First cycle int (PF)	Intensity after 20 cycles (PF) (%)	PF clusters (%)	Align (PF) (%)	Alignment score (PF)	Error rate (PF) (%)
1	789,787	277,620 ± 12,577	221,602 ± 22,142	301 ± 9	85.86 ± 4.43	79.94 ± 8.09	79.18 ± 3.10	74.65 ± 4.40	0.25 ± 0.36
2	894,690	316,300 ± 14,671	248,525 ± 7310	310 ± 8	85.59 ± 2.83	78.66 ± 2.20	80.24 ± 0.32	75.76 ± 1.44	0.25 ± 0.29
3	869,623	329,261 ± 7722	241,562 ± 38,078	290 ± 10	83.38 ± 11.53	73.46 ± 11.77	75.81 ± 11.19	71.72 ± 11.51	0.33 ± 0.84
4	907,316	318,993 ± 6080	252,033 ± 3524	272 ± 11	82.93 ± 2.27	79.03 ± 1.44	98.90 ± 0.00	160.52 ± 0.09	0.12 ± 0.00
5	782,202	264,062 ± 6363	217,279 ± 4383	269 ± 12	86.97 ± 2.42	82.30 ± 0.95	78.19 ± 0.14	75.38 ± 0.16	0.19 ± 0.01
6	829,880	388,383 ± 5665	230,522 ± 10,574	269 ± 11	84.87 ± 13.20	59.38 ± 3.05	80.95 ± 0.35	76.44 ± 0.95	0.24 ± 0.12
7	908,670	347,218 ± 12,829	252,409 ± 2606	277 ± 6	88.09 ± 2.35	72.78 ± 2.31	81.91 ± 0.15	77.99 ± 0.70	0.20 ± 0.14
8	819,793	292,730 ± 14,448	227,720 ± 7359	281 ± 18	88.53 ± 3.78	77.87 ± 2.10	79.18 ± 0.74	74.66 ± 3.06	0.42 ± 0.65

creates coordinates for single-end sequence reads:

```
bwa samse -f s_7_sequence.txt.sam
Homo_sapiens_assembly18.fasta
s_7_sequence.txt.sai
s_7_sequence.txt
```

Visualization of Alignments and File Format Conversion

SAM (Sequence Alignment/Map) is a generic human-readable tab-delimited alignment format. Although most peak-finding software accepts alignment results in SAM format, it is very common to convert SAM into a binary format known as BAM (Li et al. 2009a). BAM is also required for most visualization tools. SAMtools is an open source toolkit for SAM format file conversion as well as operations such as sorting, merging, and indexing (http://samtools.sourceforge.net/). Before converting a SAM file to a BAM file, it is useful to index the **reference sequence** in the FASTA format. This will generate a Homo_sapiens_assembly18.fasta.fai file, which will be used for the next step:

```
samtools faidx
Homo_sapiens_assembly18.fasta
```

The command to convert a SAM file to a BAM file is as follows:

```
samtools view -b -S -t
Homo_sapiens_assembly18.fasta.fai
-o s_7_sequence.txt.bam s_7_sequence.txt.sam
```

If you want to visualize the alignment results, most visualization software, such as IGV, GenomeView, and GBrowse require a sorted and indexed BAM file. The command for sorting a BAM file is as follows:

```
samtools sort s_7_sequence.txt.bam s_7_sequence.txt.sort.bam
```

This will generate a sorted s_7_sequence.txt.sort.bam file. The command for indexing a sorted BAM file is as follows:

```
samtools index s_7_sequence.txt.sort.bam
```

This will generate an indexed BAM file called s_7_sequence.txt.sort.bam.bai.

If you have both ChIP data and control (input DNA) data, you have to run all of the above steps separately for both data files.

Peak Identification

The next step is to find peaks using alignment results. From the above alignment steps, you should have two aligned, sorted, and indexed BAM files for ChIP data and control data. Now use MACS peak-finding software to find the peaks:

```
macs14 -t s_8_sequence.txt.sort.bam -c s_7_sequence.txt.sort.
bam -f BAM -g hs -nchipseq-example
```

This will generate several useful files.

1. chipseq-example_peaks.xls is the tabular file that contains information regarding the called peaks. It includes the chromosome name, peak start position, peak end position, peak length, peak summit position, total number of reads, −10*log10(pvalue), fold enrichment for this region against random Poisson distribution with local λ, and FDR in percentage.
2. chipseq-example_peaks.bed is the peak file corresponding to the peak in the chipseq-example_peaks.xls file but in BED format. This file can be uploaded to many visualization tools, such as the UCSC Genome Browser, IGV, and IGB.
3. chipseq-example_summits.bed is in BED format, which contains the peak summit locations for every peak. The fifth column in this file is the summit height of the fragment pile-up.
4. chipseq-example_negative_peaks.xls is a tabular file that contains information regarding negative peaks. Negative peaks are called by swapping the ChIP-seq and control data.
5. chipseq-example_model.r is an R script to be used to produce a PDF file containing a figure about the model based on your data.

Once the peaks are identified by MACS, it is useful to view the entire data set in the Integrative Genome Viewer (IGV) (Robinson et al. 2011). Open IGV from the Java web Start page (http://www.broadinstitute.org/igv/startingIGV). Be sure to have at least 2 GB of RAM free on your workstation. Then load BAM files for the input control and each ChIP-seq sample, and then load the corresponding **BED files** created by MACS containing the peak calls (see Figs. 2 and 3).

PEAK ANNOTATION

The fourth step is to annotate the peaks found by the peak-detection software. In standard workflow pipeline, we used a custom script to annotate each peak using UCSC's RefSeq database. The center position of each peak is used to calculate a distance to the nearest gene transcription start site (TSS) with a distance threshold. Any peak within the distance threshold to a gene's TSS will be annotated with that gene and the distance. Any peak beyond the distance threshold will be considered to be in an intergeneic region (see Fig. 9). This annotation can also be performed using the Galaxy web toolkit (Fig. 10).

Galaxy is a web-based toolkit for bioinformatic analysis of genomic intervals (http://main.g2.bx.psu.edu). It is very well suited for ChIP-seq analysis tasks such as annotation of peaks. Galaxy also includes the BWA aligner to map individual sequence reads to a reference genome and MACS for peak finding, but current NGS sequence files are so large that it is impractical to upload them to public web servers (Illumina FASTQ files may be from 4 GB to 20 GB). Most sequencing core laboratories and commercial contractors provide sequences aligned to a reference genome as a standard output file. Then it is only necessary to run MACS on a local computer in order to generate the much more compact .bed files (< 1 MB) that describe the positions of peaks.

The Cistrome web server (http://cistrome.org/ap) is a customized version of Galaxy designed for the analysis of Chip-seq and RNAseq NGS data. Use the "Import Data" command to upload a BED data file from your workstation (see Fig. 11).

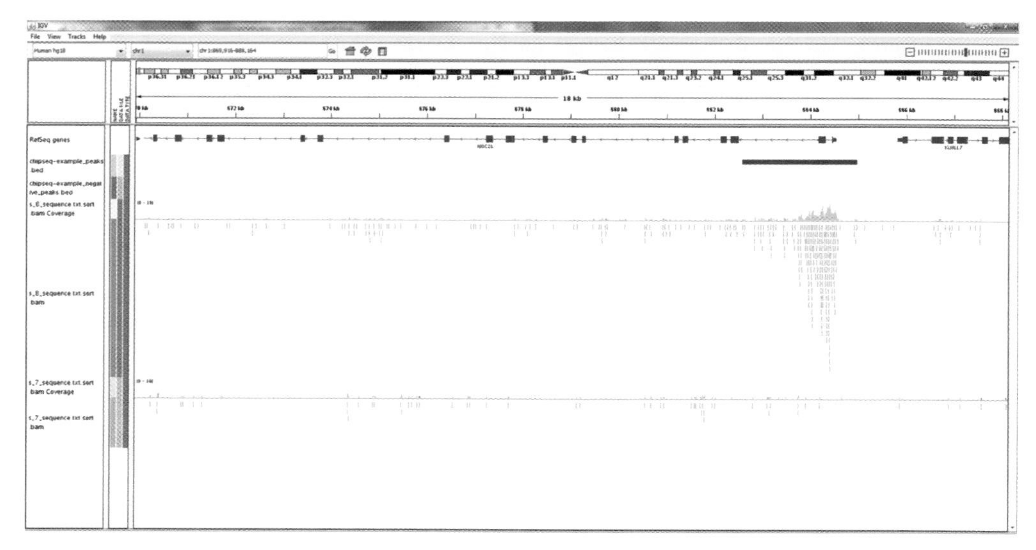

FIGURE 9. A ChIP-seq peak (MACS Peak-5) located at the TSS of the *NOC2L* gene on human chromosome 1, displayed in the Integrated Genome Viewer (Robinson et al. 2011). Many reads are shown in ChIP BAM file A, very few in the input BAM file B.

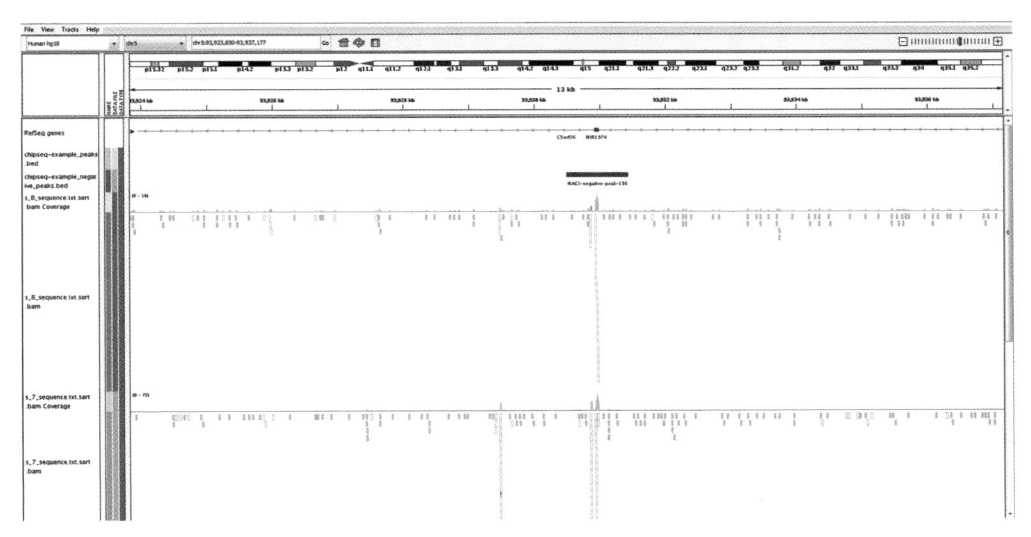

FIGURE 10. False-positive ChIP-seq peaks located in both ChIP and input (control) samples in the middle of a putative gene on human chromosome 5, displayed in the Integrated Genome Viewer (Robinson et al. 2011).

Then use the *peak2gene* annotation tool in the "Integrative Analysis" menu to map each peak to nearby genes (see Fig. 12). The tool calculates the distance (and direction) from the center of each peak to the TSS of each RefSeq gene within a specified interval (we usually use 10,000 bp).

The output of *peak2gene* is a tab-delimited file that lists the nearest gene for each peak and the distance and direction between the peak center and the TSS (see Fig. 13). Annotation for multiple genes located within the specified distance are separated by vertical lines within each column. The RefSeq annotation often lists several gene isoforms under the same gene name with the same TSS.

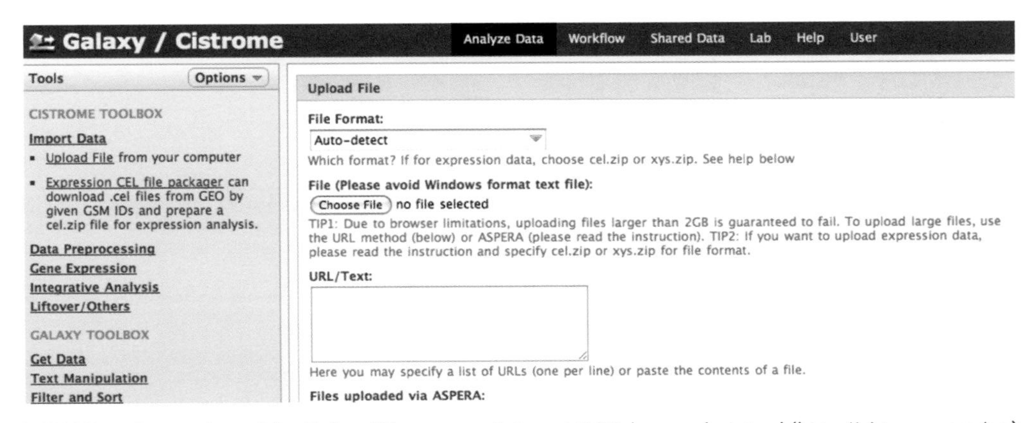

FIGURE 11. Screenshot of the Galaxy/Cistrome web-based NGS data analysis tool (http://cistrome.org/ap), file upload panel.

FIGURE 12. Screenshot of the Galaxy/Cistrome (http://cistrome.org/ap) *peak2gene* tool, used to annotate ChIP-seq peaks to the nearest gene TSS (default distance 10,000 bp).

```
Galaxy / Cistrome    Analyze Data  Workflow  Shared Data  Lab  Help  User

# Argument List:
# Name = output
# peak file = /data/CistromeAP/galaxy_database/files/000/167/dataset_167232.dat
# gene pos to peak = up,down
# distance = 5000 bp
# genome = /data/CistromeAP/static_libraries/ceaslib/GeneTable/hg18
# Output symbol as gene name = True
#chrom  pStart     pEnd       pName       pScore  NA  gene                          strand    TSS2pCenter
chr1    100087979  100089175  region_687  0       0   AGL|AGL|AGL|AGL|AGL           +|+|+|+|+     55|-350|55|55|541
chr1    10014691   10017655   region_76   0       0   UBE4B|UBE4B                   +|+       -546|-546
chr1    100207066  100209017  region_688  0       0   SLC35A3                       +         86
chr1    100274806  100275554  region_689  0       0   HIAT1                         +         1196
chr1    100275578  100276478  region_690  0       0   HIAT1                         +         348
chr1    100368979  100372569  region_691  0       0   CCDC76|SASS6                  +|-       519|-325
chr1    100486362  100488908  region_692  0       0   DBT                           -         -362
chr1    100503484  100505614  region_693  0       0   RTCD1|RTCD1                   +|+       -248|-248
chr1    10056667   10057467   region_77   0       0
chr1    100589121  100591345  region_694  0       0   CDC14A|CDC14A|CDC14A          +|+|+     377|377|377
chr1    100865171  100865960  region_695  0       0
chr1    101132751  101134167  region_696  0       0   EXTL2|SLC30A7|EXTL2|SLC30A7   -|+|-|+   453|760|453|762
chr1    101263267  101264487  region_697  0       0   DPH5|DPH5|DPH5                -|-|-     -73|-73|-73
chr1    101432422  101433113  region_698  0       0
chr1    101438613  101439296  region_699  0       0
chr1    101474548  101476205  region_700  0       0   S1PR1                         +         -484
chr1    101476698  101477383  region_701  0       0   S1PR1                         +         -2148
chr1    10192519   10193183   region_78   0       0   KIF1B|KIF1B                   +|+       499|499
chr1    10193815   10194719   region_79   0       0   KIF1B|KIF1B                   +|+       -917|-917
chr1    10380864   10381785   region_80   0       0   PGD                           +         347
chr1    10382117   10382858   region_81   0       0   PGD                           +         -816
chr1    103840949  103842259  region_702  0       0   RNPC3                         +         -439
chr1    10454165   10455760   region_82   0       0   DFFA|PEX14|DFFA               -|+|-     -238|2627|-238
chr1    10456464   10458812   region_83   0       0   DFFA|PEX14|DFFA               -|+|-     2438|-49|2438
chr1    10477328   10478353   region_84   0       0
chr1    10493126   10493964   region_85   0       0
chr1    10511209   10512087   region_86   0       0
chr1    108320011  108320668  region_703  0       0
chr1    108543124  108544070  region_704  0       0   SLC25A24                      -         -900
chr1    108544620  108545562  region_705  0       0   SLC25A24                      -         594
chr1    108904601  108905514  region_706  0       0   FAM102B                       +         -564
chr1    109035601  109037927  region_707  0       0   PRPF38B                       +         -310
chr1    109090179  109091154  region_708  0       0   STXBP3                        +         141
chr1    109091259  109091990  region_709  0       0   STXBP3                        +         -817
chr1    109160262  109161012  region_710  0       0
chr1    109174599  109175446  region_711  0       0
chr1    109221397  109222644  region_712  0       0   GPSM2                         +         -895
chr1    109419298  109420517  region_713  0       0   TAF13                         -         -240
```

FIGURE 13. Annotation output file from the Galaxy/Cistrome (http://cistrome.org/ap) *peak2gene* tool. Each peak is annotated to chromosome and start and end position, and distance to the TSS of nearby genes. Peaks with no annotation are not located within the specified distance (default 10,000 bp) of a gene TSS.

The *SeqPos* motif tool makes it very easy to search a set of ChIP-seq peaks for motifs from several databases of known transcription-factor-binding sites (Transfac, JASPAR, etc.) as well as the MDscan method for de novo discovery of enriched motifs. The standard Galaxy server (http://main.g2.bx.psu.edu) includes the EMBOSS programs, which can be used to create custom motif searches. First, use the *Fetch Sequences* tool to *Extract Genomic DNA* in FASTA format for regions identified as peaks in the .bed file created by MACS (see Fig. 14). Then use the EMBOSS fuzznuc program to search these sequences for any motif (specified using the fuzznuc regular expression syntax) with a specified number of mismatches.

There are several more sophisticated methods for motif discovery in a set of ChIP-seq peaks including Hybrid Motif Sampler (http://www.sph.umich.edu/csg/qin/HMS) (Hu et al. 2010), GimmeMotifs (http://131.174.198.125/bioinfo/gimmemotifs) (van Heeringen and Veenstra 2011), CisFinder (http://lgsun.grc.nia.nih.gov/CisFinder) (Sharov and Ko 2009), DREME (Bailey 2011), and MEME-ChIP (http://meme.nbcr.net/meme) (Machanick and Bailey 2011). See Figure 15 for a sample output of motifs found by CisFinder in a set of ChIP-seq peaks.

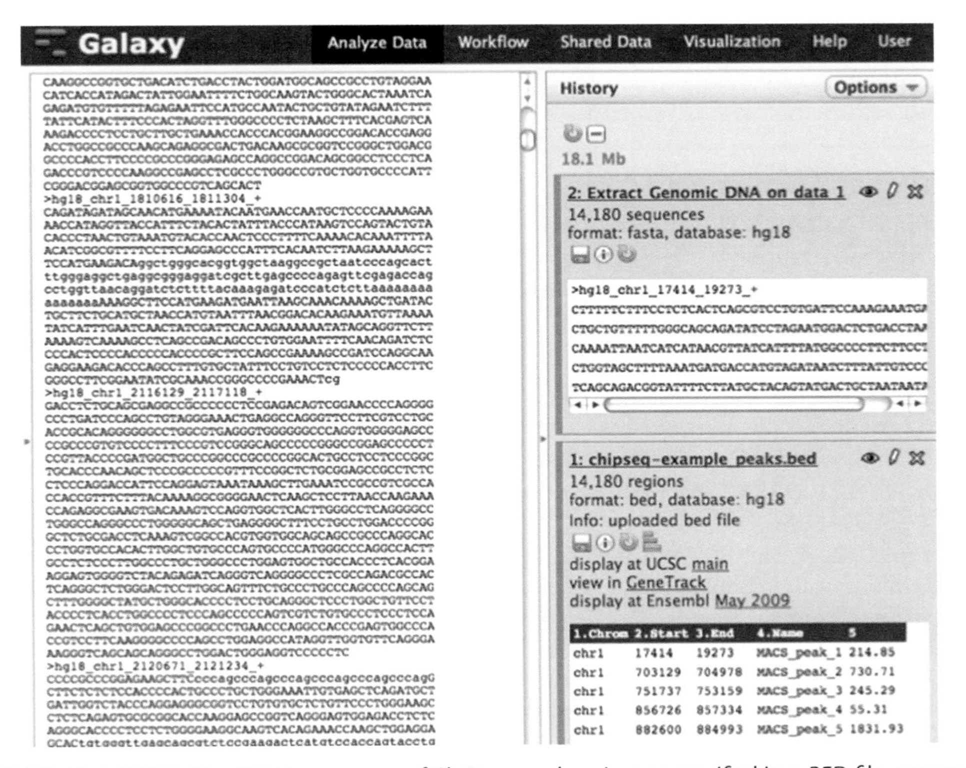

FIGURE 14. A FASTA file of DNA sequences of ChIP-seq peak regions as specified in a .BED file, extracted from the reference genome sequence by Galaxy.

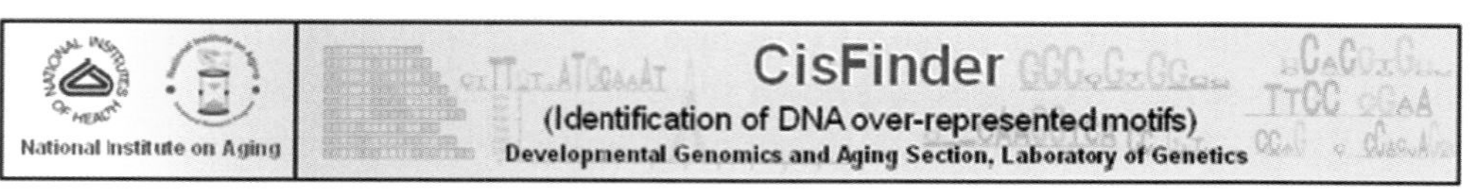

Motif M001

Freq	Ratio	Info	Score	FDR
1802	6.883	10.602	200.441	0.0000

Position Frequency Matrix (PFM)

Start position End position Reverse Name Add PFM to file

Save motif 1 14 motif_GalaxyExtractDNA

Position	A	C	G	T
1	7	45	38	8
2	4	32	18	44
3	7	20	72	0
4	10	72	15	1
5	1	18	80	0
6	0	67	32	0
7	57	24	18	0
8	0	18	24	57
9	0	32	67	0
10	0	80	18	1
11	1	15	72	10
12	0	72	20	7
13	44	18	32	4
14	8	38	45	7

FIGURE 15. Output of one highly significant motif identified by CisFinder (http://lgsun.grc.nia.nih.gov/CisFinder) (Sharov and Ko 2009) from a set of ChIP-seq peaks including a sequence logo and a position-specific scoring matrix.

REFERENCES

Bailey TL. 2011. DREME: Motif discovery in transcription factor ChIP-seq data. *Bioinformatics* 27: 1653–1659.

Barrera LO, Li Z, Smith AD, Arden KC, Cavenee WK, Zhang MQ, Green RD, Ren B. 2008. Genome-wide mapping and analysis of active promoters in mouse embryonic stem cells and adult organs. *Genome Res* **18:** 46–59.

Barski A, Zhao K. 2009. Genomic location analysis by ChIP-Seq. *J Cell Biochem* **107:** 11–18.

Barski A, Cuddapah S, Cui K, Roh TY, Schones DE, Wang Z, Wei G, Chepelev I, Zhao K. 2007. High-resolution profiling of histone methylations in the human genome. *Cell* **129:** 823–837.

Bourque G, Leong B, Vega VB, Chen X, Lee YL, Srinivasan KG, Chew JL, Ruan Y, Wei CL, Ng HH, Liu ET. 2008. Evolution of the mammalian transcription factor binding repertoire via transposable elements. *Genome Res* **18:** 1752–1762.

Dohm JC, Lottaz C, Borodina T, Himmelbauer H. 2008. Substantial biases in ultra-short read data sets from high-throughput DNA sequencing. *Nucleic Acids Res* **36:** e105. doi: 10.1093/nar/gkn425.

Fejes AP, Robertson G, Bilenky M, Varhol R, Bainbridge M, Jones SJ. 2008. FindPeaks 3.1: A tool for identifying areas of enrichment from massively parallel short-read sequencing technology. *Bioinformatics* **24:** 1729–1730.

Håndstad T, Rye MB, Drabløs F, Sætrom P. 2011. A ChIP-Seq benchmark shows that sequence conservation mainly improves detection of strong transcription factor binding sites. *PloS ONE* **6:** e18430. doi: 10.1371/journal.pone.0018430.

Hoffman BG, Jones SJ. 2009. Genome-wide identification of DNA–protein interactions using chromatin immunoprecipitation coupled with flow cell sequencing. *J Endocrinol* **201:** 1–13.

Hu M, Yu J, Taylor JM, Chinnaiyan AM, Qin ZS. 2010. On the detection and refinement of transcription factor binding sites using ChIP-seq data. *Nucleic Acids Res* **38:** 2154–2167.

Huebert DJ, Kamal M, O'Donovan A, Bernstein BE. 2006. Genome-wide analysis of histone modifications by ChIP-on-chip. *Methods* **40:** 365–369.

Ji H, Jiang H, Ma W, Johnson DS, Myers RM, Wong WH. 2008. An integrated software system for analyzing ChIP-chip and ChIP-seq data. *Nat Biotechnol* **26:** 1293–1300.

Jiang H, Wong WH. 2008. SeqMap: Mapping massive amount of oligonucleotides to the genome. *Bioinformatics* **24:** 2395–2396.

Johnson DS, Mortazavi A, Myers RM, Wold B. 2007. Genome-wide mapping of in vivo protein–DNA interactions. *Science* **316:** 1497–1502.

Jothi R, Cuddapah S, Barski A, Cui K, Zhao K. 2008. Genome-wide identification of in vivo protein–DNA binding sites from ChIP-seq data. *Nucleic Acids Res* **36:** 5221–5231.

Kharchenko PV, Tolstorukov MY, Park PJ. 2008. Design and analysis of ChIP-seq experiments for DNA-binding proteins. *Nat Biotechnol* **26:** 1351–1359.

Kim TH, Barrera LO, Ren B. 2007. ChIP-chip for genome-wide analysis of protein binding in mammalian cells. *Curr Protoc Mol Biol* **79:** 21.13.1–21.13.22.

Laajala TD, Raghav S, Tuomela S, Lahesmaa R, Aittokallio T, Elo LL. 2009. A practical comparison of methods for detecting transcription factor binding sites in ChIP-seq experiments. *BMC Genomics* **10:** 618.

Langmead B, Trapnell C, Pop M, Salzberg SL. 2009. Ultrafast and memory-efficient alignment of short DNA sequences to the human genome. *Genome Biol* ***10:*** R25. doi: 10.1186/gb-2009-10-3-r25.

Li H, Durbin R. 2009. Fast and accurate short read alignment with Burrows–Wheeler transform. *Bioinformatics* **25:** 1754–1760.

Li R, Li Y, Kristiansen K, Wang J. 2008a. SOAP: Short oligonucleotide alignment program. *Bioinformatics* **24:** 713–714.

Li H, Ruan J, Durbin R. 2008b. Mapping short DNA sequencing reads and calling variants using mapping quality scores. *Genome Res* **18:** 1851–1858.

Li H, Handsaker B, Wysoker A, Fennell T, Ruan J, Homer N, Marth G, Abecasis G, Durbin R; 1000 Genome Project Data Processing Subgroup. 2009a. The Sequence Alignment/Map format and SAMtools. *Bioinformatics* **25:** 2078–2079.

Li R, Yu C, Li Y, Lam TW, Yiu SM, Kristiansen K, Wang J. 2009b. SOAP2: An improved ultrafast tool for short read alignment. *Bioinformatics* **25:** 1966–1967.

Lin H, Zhang Z, Zhang MQ, Ma B, Li M. 2008. ZOOM! Zillions of oligos mapped. *Bioinformatics* **24:** 2431–2437.

Liu ET, Pott S, Huss M. 2010. Q&A: ChIP-seq technologies and the study of gene regulation. *BMC Biol* **8:** 56.

Machanick P, Bailey TL. 2011. MEME-ChIP: Motif analysis of large DNA datasets. *Bioinformatics* **27:** 1696–1697.

Mikkelsen TS, Ku M, Jaffe DB, Issac B, Lieberman E, Giannoukos G, Alvarez P, Brockman W, Kim TK, Bernstein BE, et al. 2007. Genome-wide maps of chromatin state in pluripotent and lineage-committed cells. *Nature* **448:** 553–560.

Mortazavi A, Williams BA, McCue K, Schaeffer L, Wold B. 2008. Mapping and quantifying mammalian transcriptomes by RNA-Seq. *Nat Methods* **5:** 621–628.

Nicol JW, Helt GA, Blanchard SG Jr, Raja A, Loraine AE. 2009. The Integrated Genome Browser: Free software for distribution and exploration of genome-scale datasets. *Bioinformatics* **25:** 2730–2731.

Nix DA, Courdy SJ, Boucher KM. 2008. Empirical methods for controlling false positives and estimating confidence in ChIP-seq peaks. *BMC Bioinformatics* **9:** 523.

Park PJ. 2009. ChIP-seq: Advantages and challenges of a maturing technology. *Nat Rev Genet* **10:** 669–680.

Pepke S, Wold B, Mortazavi A. 2009. Computation for ChIP-seq and RNA-seq studies. *Nat Methods* **6:** S22–S32.

Quail MA, Kozarewa I, Smith F, Scally A, Stephens PJ, Durbin R, Swerdlow H, Turner DJ. 2008. A large genome center's improvements to the Illumina sequencing system. *Nat Methods* **5:** 1005–1010.

Robertson G, Hirst M, Bainbridge M, Bilenky M, Zhao Y, Zeng T, Euskirchen G, Bernier B, Varhol R, Delaney A, et al. 2007. Genome-wide profiles of STAT1 DNA association using chromatin immunoprecipitation and massively parallel sequencing. *Nat Methods* **4:** 651–657.

Robinson JT, Thorvaldsdóttir H, Winckler W, Guttman M, Lander ES, Getz G, Mesirov JP. 2011. Integrative genomics viewer. *Nat Biotechnol* **29:** 24–26.

Rozowsky J, Euskirchen G, Auerbach RK, Zhang ZD, Gibson T, Bjornson R, Carriero N, Snyder M, Gerstein MB. 2009. PeakSeq enables systematic scoring of ChIP-seq experiments relative to controls. *Nat Biotechnol* **27:** 66–75.

Sharov AA, Ko MS. 2009. Exhaustive search for over-represented DNA sequence motifs with Cis-Finder. *DNA Res* **16:** 261–273.

Smith AD, Xuan Z, Zhang MQ. 2008. Using quality scores and longer reads improves accuracy of Solexa read mapping. *BMC Bioinformatics* **9:** 128.

Smith AD, Chung WY, Hodges E, Kendall J, Hannon G, Hicks J, Xuan Z, Zhang MQ. 2009. Updates to the RMAP short-read mapping software. *Bioinformatics* **25:** 2841–2842.

Solomon MJ, Larsen PL, Varshavsky A. 1988. Mapping protein–DNA interactions in vivo with formaldehyde: Evidence that histone H4 is retained on a highly transcribed gene. *Cell* **53:** 937–947.

Stein LD, Mungall C, Shu S, Caudy M, Mangone M, Day A, Nickerson E, Stajich JE, Harris TW, Arva A, Lewis S. 2002. The generic genome browser: A building block for a model organism system database. *Genome Res* **12:** 1599–1610.

Valouev A, Johnson DS, Sundquist A, Medina C, Anton E, Batzoglou S, Myers RM, Sidow A. 2008. Genome-wide analysis of transcription factor binding sites based on ChIP-seq data. *Nat Methods* **5:** 829–834.

van Heeringen SJ, Veenstra GJ. 2011. GimmeMotifs: A de novo motif prediction pipeline for ChIP-sequencing experiments. *Bioinformatics* **27:** 270–271.

Vega VB, Cheung E, Palanisamy N, Sung WK. 2009. Inherent signals in sequencing-based chromatin-immunoprecipitation control libraries. *PloS ONE* **4:** e5241. doi: 10.1371/journal.pone.0005241.

Wilbanks EG, Facciotti MT. 2010. Evaluation of algorithm performance in ChIP-seq peak detection. *PloS ONE* **5:** e11471. doi: 10.1371/journal.pone.0011471.

Zhang Y, Liu T, Meyer CA, Eeckhoute J, Johnson DS, Bernstein BE, Nusbaum C, Myers RM, Brown M, Li W, Liu XS. 2008. Model-based analysis of ChIP-Seq (MACS). *Genome Biol* **9:** R137. doi: 10.1186/gb-2008-9-9-r137.

WWW RESOURCES

http://cistrome.org/ap The Cistrome web server is a customized version of Galaxy designed for the analysis of Chip-seq and RNA-seq NGS data.

http://genomeview.org/ GenomeView is a genome browser and editor developed at Broad Institute. © Thomas Abeel.

http://hgdownload.cse.ucsc.edu/downloads.html This page contains links to sequence and annotation data downloads for the genome assemblies featured in the UCSC Genome Browser.

http://info.gersteinlab.org/PeakSeq The PeakSeq software (described in Rozowsky et al. 2009) is a program for identifying and ranking peak regions in ChIP-seq experiments.

http://lgsun.grc.nia.nih.gov/CisFinder CisFinder is a tool for finding over-representing short DNA motifs (e.g., transcription factor binding motifs) and estimates the Position Frequency Matrix (PFM) directly from word counts. It is designed for the analysis of ChIP-chip and ChIP-seq data and can process long input files (50 Mb).

http://main.g2.bx.psu.edu Galaxy an open, web-based platform for data-intensive biomedical research.

http://meme.nbcr.net/meme The MEME homepage.

http://novocraft.com The Novocraft Technologies home page.

http://samtools.sourceforge.net/ The SAM (Sequence Alignment/Map) format is a generic format for storing large nucleotide sequence alignments

https://sourceforge.net/projects/chipseqtools/files The TRLocator software for ChIP-seq peak detection, developed by CHIBI at NYU.

http://www.biostat.jhsph.edu/~hji/cisgenome An integrated tool for tiling array, ChIP-seq, genome, and *cis*-regulatory element analysis). © The Johns Hopkins University.

http://www.broadinstitute.org/igv/startingIGV The Integrative Genomics Viewer home page, Broad Institute.

http://www.ncmls.eu/bioinfo/gimmemotifs GimmeMotifs is a de novo motif predication pipeline, especially suited for ChIP-seq data sets (see van Heeringen and Veenstra 2011).

http://www.sph.umich.edu/csg/qin/HMS HMS (hybrid motif sampler) implements a novel computational algorithm specifically designed for transcription factor binding sites (TFBS) motif discovery using ChIP-seq data. Center for Statistical Genetics, University of Michigan.

10

RNA Sequencing with NGS

Stuart M. Brown, Jeremy Goecks, and James Taylor

Sequencing of RNA has been an important application of DNA sequencing technology since its invention. RNA is usually sequenced by first converting it to complementary DNA (cDNA) with the reverse transcriptase enzyme (RNA-dependent DNA polymerase). Reverse transcriptase was originally isolated from Rous sarcoma retrovirus and Rauscher mouse leukemia retrovirus (R-MLV) by David Baltimore (1970) and independently by Howard Temin (Temin and Mizutani 1970). In 1972, Verma et al. (1972) and Bank et al. (1972) developed efficient systems to copy messenger RNA (mRNA) to cDNA by adding DNA nucleotide triphosphates and oligo(dT), which hybridizes to the poly(A) tail of the mRNAs and acts as a primer.

cDNA is frequently the subject of sequencing studies because this an efficient method to discover the coding sequence of expressed genes or for finding gene-coding regions. Craig Venter expanded this method by collecting large numbers of short single reads from the 3′ ends of mRNA, which were called expressed sequence tags (ESTs). Early EST sequencing of human cells was extraordinarily productive, resulting in the discovery of many thousands of new genes (Adams et al. 1991, 1992). The EST method allowed for a rough form of gene expression measurements in a variety of cell types, and some differential expression studies were conducted in this manner. EST sequencing also became a valuable component of **de novo sequencing** projects, providing a layer of expression information to annotation and gene-finding efforts. **Microarray** technology, developed in the 1990s, measures the hybridization of labeled cDNA to an array of DNA probes that correspond to the sequences of known genes. The microarray method allows for the discovery, in a genome-wide fashion, of gene expression changes (as reflected in changes of mRNA levels) resulting from any biological treatment or condition.

RNA sequencing with NGS technology can be used for several different scientific applications. Direct sequencing of mRNA provides a measurement of gene expression for the entire transcriptome that is more accurate and has a greater dynamic range than microarray-based technologies (Marioni et al. 2008). **RNA-seq** can also be used to detect mutations in transcribed portions of the genome for the native germline cells of an individual or for somatic mutations in tumor cells. RNA-seq is also an excellent platform to measure alternative splicing events that produce different transcripts (and ultimately different proteins) from a single gene. Alternative transcript isoforms can be detected with great accuracy by using RNA-seq reads mapping at splice junctions, specifying both known as well as novel isoforms. With appropriate sample preparation methods, RNA-seq can also be used to interrogate a wide variety of non-protein-coding RNAs.

Protocols for sequencing of RNA have been developed by all of the major NGS vendors. **Ribosomal RNA** (rRNA) and transfer RNA (tRNA) are very abundant in the total RNA extracted from both prokaryotic and eukaryotic cells (~75% of RNA molecules). Sequencing of abundant non-protein-coding RNA reduces the yield and sensitivity of RNA-seq methods for mRNA and increases cost. Most protocols for RNA-seq in eukaryotic cells use poly(T) oligonucleotides to isolate mRNA with poly(A) tails or use poly(T) primers in combination with random short oligomers for reverse transcription. After poly(A) purification, most protocols shatter mRNA molecules into small fragments (from 100 to 300 bp), which are then ligated with oligomers specific for the sequencing system. Some protocols have also been developed to sequence small non-protein-coding RNA molecules such as microRNA (miRNA), small interfering RNA (siRNA), small nuclear RNA (snRNA), small nucleolar RNA (snoRNA), Piwi-interacting RNA (piRNA), and others.

Another method of removing abundant rRNA, tRNA, and other highly abundant RNAs before sequencing is called duplex-specific nuclease (DSN) normalization. This method uses a nuclease (Kamchatka crab hepatopancrease) that specifically degrades double-stranded DNA while leaving single-stranded DNA molecules intact. This method takes advantage of reassociation kinetics (Zhulidov et al. 2004). First, total RNA is reverse-transcribed to double-stranded cDNA. Then the cDNA is denatured at high temperature. Under selective annealing conditions, the most abundant cDNA molecules (cDNA clones of rRNA, tRNA, mtRNA, and the most highly transcribed messages) form double strands and are degraded while less abundant molecules remain single stranded and are preserved. Illumina has presented some preliminary data using DSN normalization for RNA-seq (http://www.illumina.com/documents/seminars/presentations/2010-06_sq_03_lakdawalla_transcriptome_sequencing.pdf) that indicate very good removal of rRNA, high retention of small non-coding RNAs, and no 3′ bias compared with poly(A) purification methods. Relatively few RNA-seq experiments have yet been published using the DSN method, thus it is not yet clear what bias it creates in gene expression or differential expression values.

Despite purification methods (poly(A) selection or DSN normalization), RNA-seq data may contain substantial amounts of rRNA, tRNA, and also mitochondrial RNA. These can be filtered out in the bioinformatics pipeline by simply using a "contaminant" file of rRNA, tRNA, and mtDNA sequences for the target species to prefilter all **sequence reads** (by **alignment**) before mapping the remaining reads to the genome and/or splice junction database.

MEASURING GENE EXPRESSION

After sequencing, the standard approach to RNA-seq data analysis requires mapping of all NGS reads to a **reference genome**, using **sequence alignment** software, and measuring expression for each gene by counting the number of sequence reads that align to its coding region. Alignment of millions of short RNA-seq reads is generally performed using either a software method based on the **Burrows–Wheeler transformation**, such as BWA (Li and Durbin 2009) or Bowtie (Langmead et al. 2009), or a method based on a hash table of short ***k*-mers** (strings of sequence letters) such as SHRiMP (Rumble et al. 2009), BFAST (Homer et al. 2009), and the Illumina ELAND aligner (see Chapter 4). All alignment methods find perfect matches, matches with one or two mismatches, and very small insertion/deletions and filter out repeated sequences that match the genome in many different locations. Alignment is generally the most computationally demanding step in any NGS analysis workflow, requiring many hours on a cluster of **high-performance computers** (see Chapter 12).

There are informatics challenges in aligning short NGS RNA reads as query sequences to a genomic DNA reference, because the RNA sequences have **introns** spliced out creating gaps in the alignment. Additional alignment problems that are common to NGS of both genomic DNA and RNA include SNPs (and other **sequence variants**) between the actual biological source of the RNA being sequenced and the reference genome, sequencing errors, and reads that align to multiple positions on the genome. RNA editing is a newly discovered source of differences between sequenced cDNA fragments and the reference genome (Li et al. 2011).

Possible approaches to the intron splicing problem include building a reference database of transcript sequences, using an alignment method tolerant of large gaps, or creating a supplemental database of predicted splice junction fragments with flanking **exon** sequences. If the reference database (i.e., RefSeq gene annotations) includes alternative transcripts for each gene, the quantitation of gene expression becomes even more complex. Mapping only to known transcripts undermines the power of RNA-seq to find novel splice isoforms and expression from unannotated portions of the genome (new genes or new exons of known genes).

Illumina has developed a software method called ELAND/CASAVA to quantify gene expression for its sequencers. A reference genome, annotated with the locations of exons and introns for each gene, is processed into a splice junction database. The junction sequences are extracted with flanking exon regions several bases shorter than the length of the RNA-seq reads, so that each read is required to straddle the splice junction and be anchored to both exons. If a gene has multiple isoforms, the coding region is considered to be the union of all regions annotated as exons in any isoforms, and multiple incompatible splice junction fragments may exist in the junction database. This method allows for correct mapping of most RNA-seq reads to their corresponding gene, regardless of the placement of the read with respect to splice junctions. The Illumina method calculates the gene expression value as the sum of the total number of bases from all reads that align to exon and splice junction locations for each gene. The CASAVA software also reports the number of reads that map to each exon, reads that map entirely to introns, or reads that map outside of the annotated exonic region (3′ and 5′ extensions of the gene). Users of Illumina software should be aware that the ELAND/CASAVA method has undergone so many changes in various software versions that RNA-seq gene expression values may change depending on the software version used to calculate them.

The TopHat/Cufflinks software package (Trapnell et al. 2009) takes a different approach to gene expression and intron splicing. Rather than rely on a database of known transcript isoforms, TopHat identifies splice junctions de novo from gaps in the alignment of reads to the genome. Cufflinks can then quantify the sum of all reads that map to a gene across exons and all intron splice junctions (Trapnell et al. 2010). Accurate de novo discovery of splice junctions requires very deep sequencing **coverage**, because each splice site must be covered by multiple reads with adequate overlap with the exons on both sides to allow for alignment.

Important factors that influence the quality and interpretability of RNA-seq data are the total yield of sequence data and coverage across the transcriptome. It is well known that genes are transcribed at different levels in different cell types and under different conditions in accordance with the needs of the cell for gene expression. The abundance of transcripts from different genes observed in RNA-seq data has been shown to accurately represent the gene expression profile of various cell samples when validated by other technologies such as RNA microarray and quantitative PCR (Marioni et al. 2008). As the total yield of NGS machines has increased, the sensitivity of RNA-seq has greatly exceeded microarray-based methods of measuring transcripts from genes expressed at low levels. Because RNA-seq does not rely on existing sequence data for the creation of probes, it can measure the expression of unannotated genes and portions of known genes not previously observed in transcripts such as 5′ and 3′ extensions as well as a variety of alternatively spliced isoforms that include regions annotated as introns.

Pickrell et al. (2010) found that ~15% of mapped RNA-seq reads were located outside of annotated exons. Figure 1 illustrates the RNA-seq mapping around the *ADM* gene, with large numbers of reads mapping to annotated exons, but some reads also mapping to introns and 5′ regions.

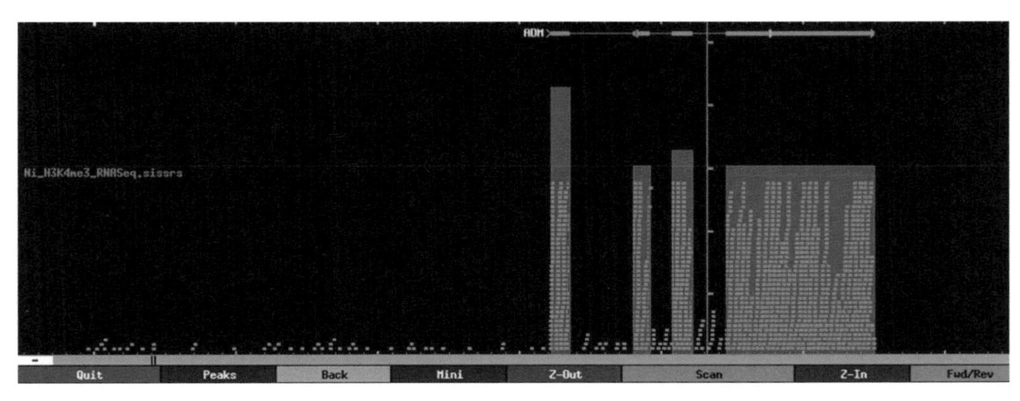

FIGURE 1. RNA-seq reads mapped to the human genome in the region of the *ADM* gene. (Image courtesy of P.R. Smith, New York University.)

NORMALIZATION AND DIFFERENTIAL EXPRESSION

The fundamental design of gene expression experiments for RNA-seq is very similar to microarrays. The most common goal of RNA-seq experiments is to measure changes in genome-wide gene expression (differential expression). Statistical analysis is performed to identify individual genes (and functional groups of genes) that have changed significantly in abundance across experimental conditions. Although the technical reproducibility of RNA-seq is very high, the underlying variability of abundance of RNA molecules in biological samples is high, thus it is necessary to use replicate samples to study changes in gene expression across experimental treatments.

Although the sensitivity and specificity of RNA-seq are very good, there are some informatics issues associated with accurate quantitation of gene expression. Longer mRNA transcripts receive more reads because they contribute more fragments to the sequencing library. Sequence-specific bias may also occur in various steps of the sample prep, sequencing, and alignment. Fortunately, many of these biases cancel out when calculated in differential expression between two different biological conditions for the same set of genes. However, sample prep and machine-specific sequencing issues often lead to differences in the total yield of sequence reads among a set of samples for a biological experiment. This creates a need for normalization methods across samples in order to make accurate comparisons and calculate differential expression scores for each gene. The simple reads per kilobase per million (RPKM) normalization method developed by the Wold laboratory at Caltech (Mortazavi et al. 2008) has become the de facto standard. RPKM normalization divides the read count for each gene by the length of the RefSeq transcript for that gene and then scales all read counts per million reads in each lane of sequencing. Without careful normalization, longer genes are more likely to be identified as differentially expressed. More recent statistical studies indicate that a few very highly expressed genes may create bias in the RPKM values, thus a quantile-based normalization method will provide a more accurate measurement of differential expression for low-copy-number genes (Bullard et al. 2010).

Because most RNA-seq sample preparation protocols include purification by binding to poly(T) or priming of reverse transcriptase with poly(T) oligos, a 3′ bias may be present in the distribution of sequence reads across transcripts. This may not have a large effect on calculations of differential gene expression (assuming the 3′ bias is consistent for all samples and all genes in an experiment), but it can have a dramatic effect on measurements of alternative splicing, because 5′ regions of longer genes may have low coverage and high variability across samples. If 3′ bias differs across samples in a single experiment, it may increase the measured gene expression values for shorter genes in lanes with greater 3′ bias.

RNA-seq expression data can be calculated for individual exons. These data can then be further analyzed to identify exons that show changes in expression across biological conditions that differ from the changes in expression levels of the entire gene (exon-specific expression).

ALTERNATIVE SPLICING

The detection and quantitation of alternative splicing events in an RNA-seq data set is a challenging bioinformatics problem. The CASAVA method of mapping reads to exons relies on a database of "non-overlapping" exons that combines overlapping exons of different length (alternative 3′ or 5′ splice sites) and ignores situations in which an entire intron is alternatively spliced or retained. In fact, one of the most challenging aspects of studying alternative splicing is the definition of the alternative splice events and building a database of valid alternatively spliced transcripts or isoforms. Wang et al. (2008) defined eight types of alternative transcription events: skipped exon, retained intron, alternative 5′-splice site, alternative 3′-splice site, mutually exclusive exons, alternative first exon, alternative last exon, and tandem UTRs. We also frequently observe the use of additional first exons and additional last exons. Any method of quantifying alternative splicing must first define the set of alternative transcripts possible for each gene. Methods that rely on a database of known transcript annotations are very dependent on the database used. The RefSeq database is very stringent in its definition of alternative transcripts and typically only includes four to five transcripts per gene. In contrast, the AceView database (Thierry-Mieg and Thierry-Mieg 2006) shows evidence for many more splice variants per gene (Fig. 2).

The analysis of alternative splicing in RNA-seq data presents major challenges in transcript assembly and abundance estimation, arising from the ambiguous assignment of reads to isoforms. Because alternatively spliced transcripts for a single gene often share many exons, reads that map to shared exons cannot be assigned to one particular isoform (see Fig. 3). Only reads that span a specific, previously annotated, splice junction

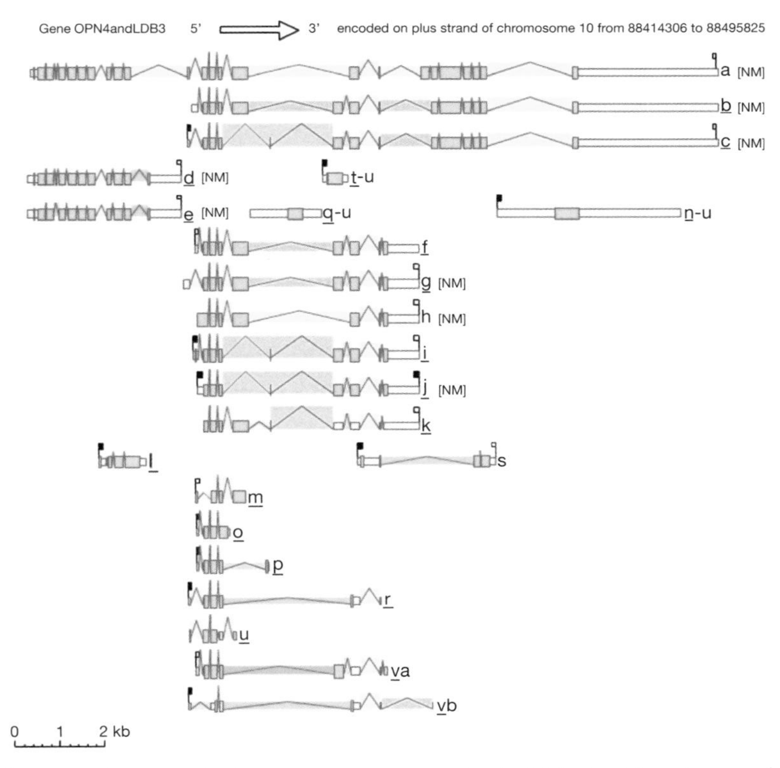

FIGURE 2. The gene *OPN4* has 23 splice isoforms annotated in the NCBI AceView database (http://www.ncbi.nlm.nih.gov/IEB/Research/Acembly/index.html?human).

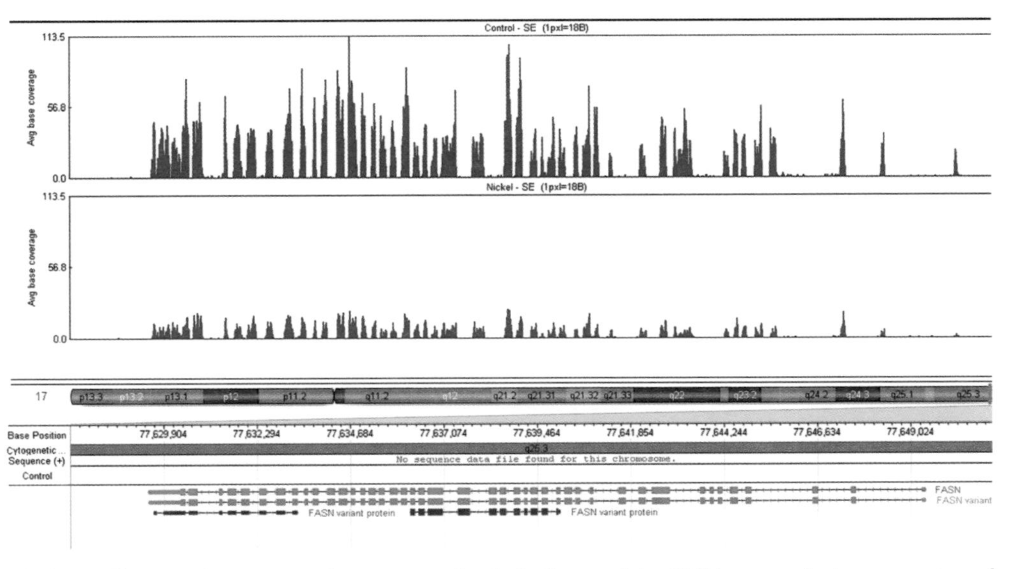

FIGURE 3. The complex pattern of exon expression in isoforms of the *FASN* gene make interpretation of differential splicing across samples/treatments difficult. (Image courtesy of Z. Tang, New York University.)

or reads that map uniquely to just one isoform (i.e., a cassette exon) are unambiguous. Several software methods have been developed to analyze RNA-seq data that are intended to provide estimates of alternative splicing events. The Cufflinks program, developed by Cole Trapnell and Steven Salzberg at the University of Maryland in collaboration with members of the Wold laboratory at Caltech and others (Trapnell et al. 2010), builds sets of synthetic alternative transcripts directly from the genome mapped RNA-seq reads, and then quantitates the abundance of each of these transcripts, without relying on a database of exon/intron annotations. Cufflinks was designed to work with the Bowtie and TopHat sequence alignment programs, but it can use any file of RNA-seq reads aligned to a reference genome in SAM or BAM format, such as that produced by the BWA aligner or the Illumina ELAND/CASAVA package. Cufflinks works best with **paired-end reads**, where it can calculate the size of the RNA fragments used for sequencing, which, in turn, can provide a great deal of additional data regarding splice isoforms (a change in distance between the genome mapped position of paired-end reads on an mRNA represents a change in splice isoform). Cufflinks includes the program Cuffdiff, which can find significant changes in transcript expression, splicing, and promoter use (i.e., alternative transcription start sites). Cufflinks requires deep coverage of the transcriptome without extensive 3′ bias in order to work well.

Current methods of detection for alternative splicing can provide a good picture of genome-wide splicing activity in particular cell types and can be used to test for changes in isoform abundance for specific genes across samples. However, these methods tend to have a high false discovery rate when used to find any genome-wide changes in isoform use across a set of biological conditions.

DOWNSTREAM ANALYSIS

Once genes have been identified as differentially expressed (or alternatively spliced) by appropriate statistical tests of RNA-seq data, downstream analysis can proceed using similar tools as for microarray gene expression studies. Lists of differentially regulated genes can be analyzed for functional correlation by enrichment of GO terms using hypergeometric tests or by gene set enrichment. These tools may need some adjustment as specific biases inherent in RNA-seq versus microarrays become more clearly defined, such as a correction for transcript length.

There is also great potential for integration of RNA-seq data with other kinds of count-based genomic data that can be derived from NGS platforms. In particular, integration of RNA-seq gene expression measurements with **ChIP-seq** measurements of transcription factor binding and/or **histone** modification patterns can provide a deeper and more nuanced picture of gene expression activity. When coupled with genomic sequencing (or genotyping), RNA-seq can be used to measure **allele**-specific expression (see Fig. 4).

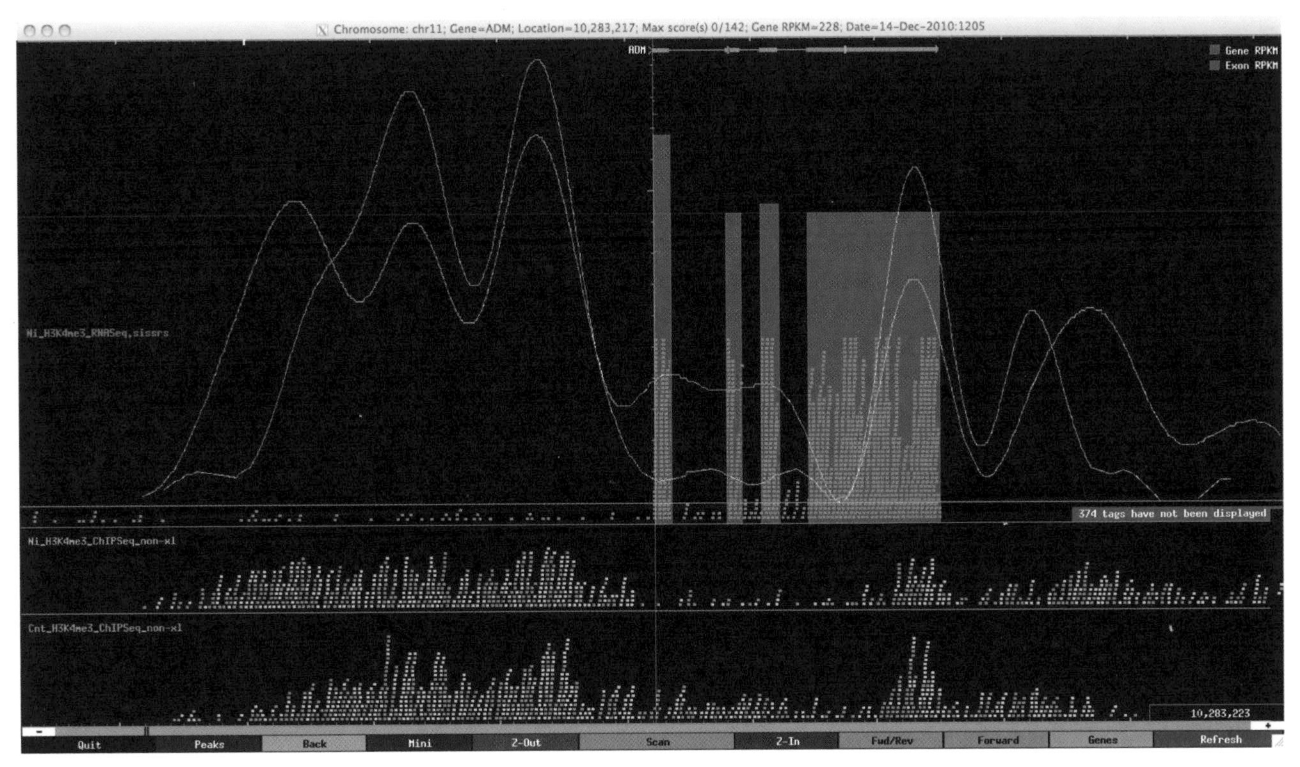

FIGURE 4. A combined view of RNA-seq and ChIP-seq data illustrates the effects of histone modification on gene expression and the interaction of histones with RNA polymerase. (Image courtesy of P.R. Smith, New York University.)

TUTORIAL: GALAXY RNA-seq ANALYSIS EXERCISE

RNA-seq analysis is a challenging computational problem—both in terms of the complexity of the **algorithms** and the usability of the current generation of software, as well as the requirements for high-powered computers to map and summarize millions of reads across genes, exons, transcripts, and so on. There are some commercial software packages that simplify the task of calculating RNA-seq gene expression from **FASTQ files**. Many core laboratories and sequencing vendors offer standard RNA-seq data analysis pipelines either included in the cost of sequencing or for a standard analysis fee. The Galaxy Web Service offers a simplified interface to the TopHat/Cufflinks RNA-seq analysis toolkit; however, it takes a substantial amount of time to upload and execute these computations on the free version of Galaxy. Anyone planning on analyzing a substantial amount of RNA-seq samples will either need to access a powerful UNIX server where they will need to install and run TopHat/Cufflinks (or other similar software), install a local copy of Galaxy to use the simplified interface with high-performance computing power (work with your local IT on this one), or use a cloud-based service in which computing power can be rented by the hour on a system with preinstalled tools (see Chapter 12). The Galaxy Cloud Project (http://wiki.g2.bx.psu.edu/Admin/Cloud) makes it very easy to rent access to your own private copy of Galaxy set up for you on the Amazon Elastic Compute Cloud (EC2).

The Galaxy Project team has created a tutorial for RNA-seq analysis that includes QC, trimming of low-quality portions of sequence reads, mapping of reads to a reference genome, quantifying reads per transcript, and comparison of two samples to find differential gene expression (Goecks and Taylor 2011). This is a time-consuming tutorial, even though the training data are already uploaded and stored on the Galaxy server and the training data set represents only a tiny fraction of an actual NGS RNA-seq run. Once again, realize that it is not feasible to analyze substantial amounts of data using this free server.

See http://main.g2.bx.psu.edu/u/jeremy/p/galaxy-rna-seq-analysis-exercise. Galaxy provides multiple tools for performing RNA-seq analysis. This exercise introduces these tools and guides use of these tools on some example data sets; prominent RNA-seq tools include TopHat and Cufflinks. Familiarity with Galaxy and the general concepts of RNA-seq analysis are useful for understanding this exercise. This exercise should take 1–2 h.

Below are small samples of data sets from the ENCODE Caltech RNA-seq track; specifically, the data sets are single 75-bp reads from the h1-hESC and GM12878 cell lines (see Fig. 5).The sampled reads map mostly to Chr19. Import the data sets to your history by clicking on the green-plus icon labeled "Import."

http://main.g2.bx.psu.edu/datasets/7f717288ba4277c6/imp

http://main.g2.bx.psu.edu/datasets/257ca40a619a8591/imp

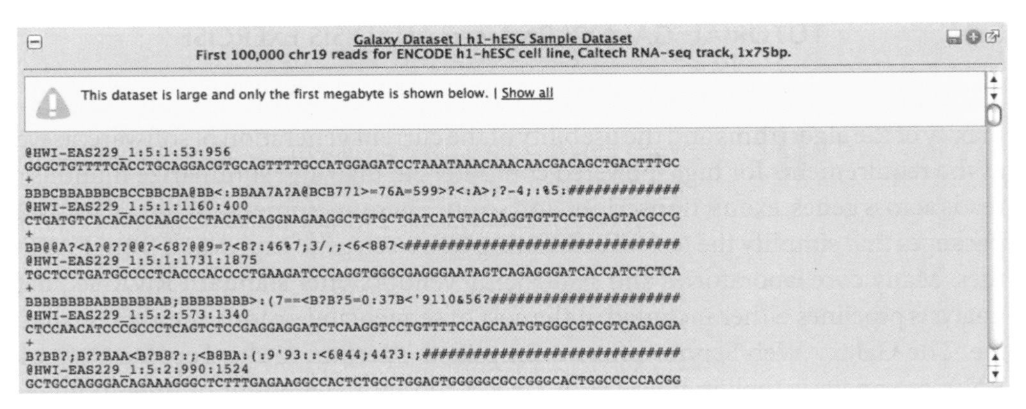

FIGURE 5. Sample lines from RNA-seq sample data set ENCODE h1-hESC with 75-bp reads (see https://main.g2.bx.psu.edu/).

Understanding and Preprocessing the Reads

You should understand the reads a bit before analyzing them. Preprocessing may be needed as well.

Step 1: Compute statistics and create a boxplot of base-pair quality scores for each set of reads using the [`NGS: QC and manipulation` ➢] `FASTQ Summary Statistics` tool and then plot the output using the [`Graph/Display Data` ➢] `Boxplot`. Often, it is useful to trim reads to remove base positions that have a low median (or bottom quartile) score. For this exercise, assume a median quality score below 15 to be unusable (see Fig. 6). Given this criterion, is trimming needed for the data sets? If so, which base pairs should be trimmed?

Step 2: If necessary, trim the reads based on your answers to Step 1 using [`NGS: QC and manipulation` ➢] `FASTQ Trimmer`.

Map Processed Reads

The next step is mapping the processed reads to the genome. The major challenge when mapping RNA-seq reads is that the reads, because they come from RNA, often cross splice junction boundaries; splice junctions are not present in a genome's sequence, and hence typical NGS mappers such as Bowtie and BWA are not ideal without modifying the genome sequence. Instead, it is better to use a mapper such as TopHat that is designed to map RNA-seq reads.

Step 1: Use the [`NGS: RNA Analysis` ➢] `TopHat` tool to map RNA-seq reads to the hg18 genome build. This will take ~10 min. Look at the documentation to understand the two data sets that TopHat produces. How many splice junctions did TopHat find? Are most splice junctions supported by (i.e., spanned by) more

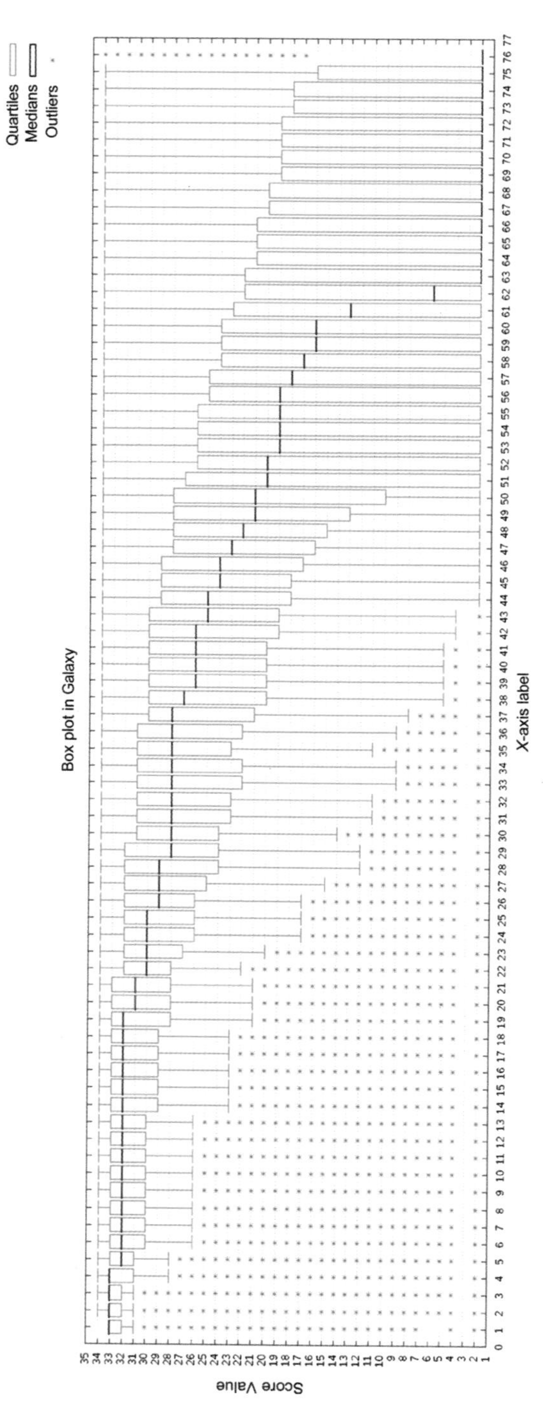

FIGURE 6. Galaxy boxplot of sequence quality scores showing severe decline in quality at the 3′ end of the reads (see https://main.g2.bx.psu.edu/).

or less than 10 reads? (Hint: the score column in the splice junctions data set is useful for answering this question.)

Step 2: To view TopHat's output, create a simple Galaxy visualization by selecting `Visualization` ➢ `New Track Browser` from the main Galaxy menu at the top. After creating the visualization, add data sets to your visualization by clicking on the `Options` ➢ `Add Tracks` option. Add the "accepted hits" BAM data set and the "splice junctions" data set produced by TopHat, and also the UCSC human Reference Genes (refGene chr19) for chromosome 19 (you will need to import the data set before you can add it to your visualization):

http://main.g2.bx.psu.edu/datasets/965c374b65239597/imp

Navigate to chr19 using the select box at the top of visualization and look at the data. This visualization makes clear how little of the data you are working with. Zoom in to view the data at the beginning of the chromosome. You can zoom in by (1) double-clicking anywhere on the visualization to zoom in on that area or (2) dragging on the base number area at the top of the visualization to create a zoom area. You should be able to see (1) the reads mapped by TopHat, including splice junctions (reads that are split by a splice junction are connected by a thin line in the visualization); (2) just the splice junctions produced by TopHat; and (3) how TopHat's reads and junctions correspond to UCSC's RefSeq gene track. Find an example of a splice junction between two known exons, and find an example where a splice junction should be found but is not (see Fig. 7).

Assemble and Analyze Transcripts

After mapping the reads, the next step is to assemble the reads into complete transcripts that can be analyzed for differential expression and phenomena such as splicing events and transcriptional start sites.

Step 1: Run [`NGS: RNA Analysis` ➢] `Cufflinks` on each BAM data set produced by TopHat to perform de novo transcript assembly. Check the documentation for the data sets that Cufflinks produces. How can you tell when a transcript has multiple exons?

Step 2: Run [`NGS: RNA Analysis` ➢] `Cuffcompare` on the assembled transcripts and use the UCSC RefSeq Genes data set as the reference annotation. Cuffcompare requires that the reference annotation be in GTF format, thus use this version of the UCSC RefSeq Genes. Find some transcripts that appear in both samples and have FPKM confidence bands that do not overlap. You can find this information by looking at the "transcript tracking" data set produced by Cuffcompare and reading the Cuffcompare documentation to understand the data in this data set.

Step 3: Add the Cufflinks' assembled transcripts data sets to the visualization you created earlier in order to view the transcripts alongside the mapped reads,

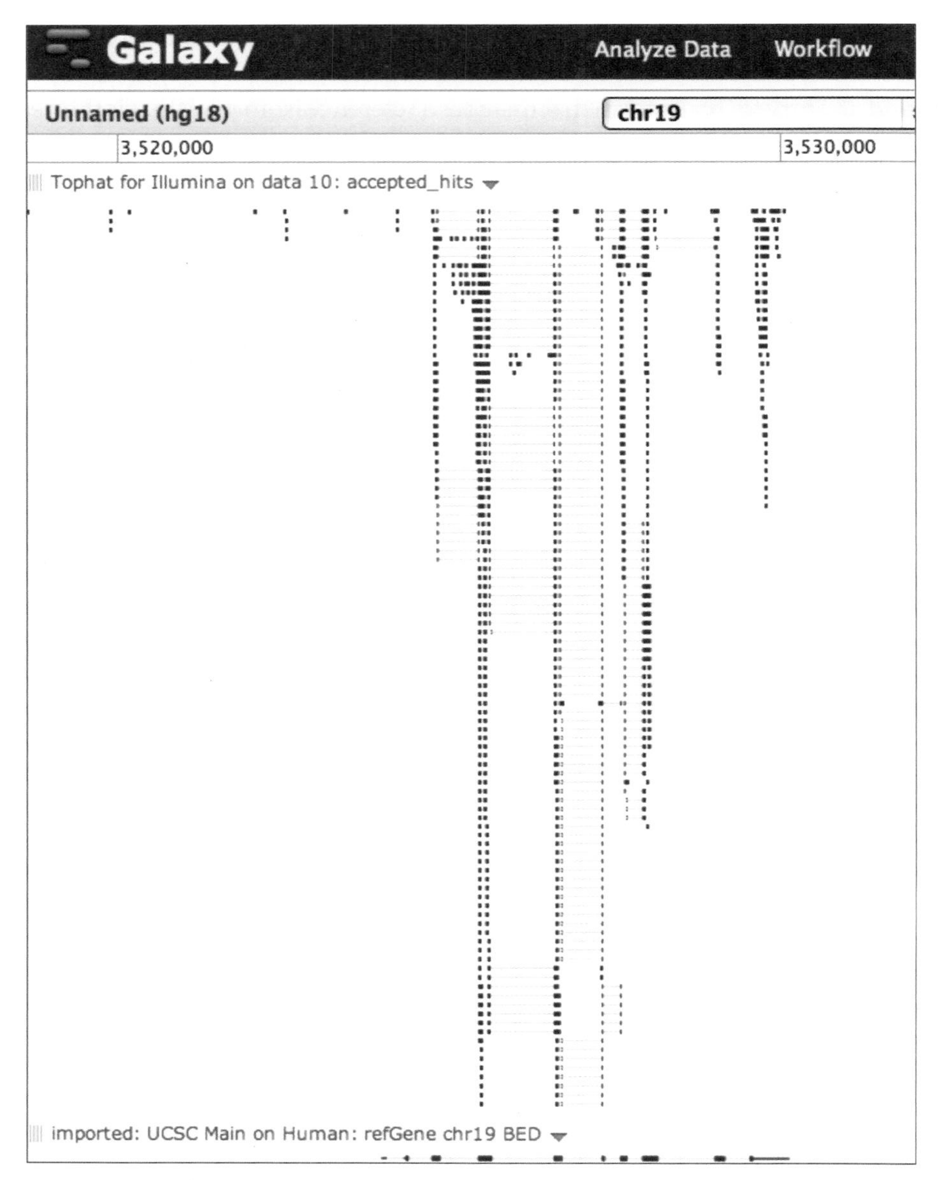

FIGURE 7. Galaxy visualization of RNA-seq reads aligned by TopHat across splice junctions of RefSeq gene NM_006339 on human hg18 build of chromosome 19 (see https://main.g2.bx.psu.edu/).

junctions, and reference genes. Can you find examples where Cufflinks/Cuffcompare assembled a complete or almost complete transcript?

Step 4: Run [`NGS: RNA Analysis` ➢] `Cuffdiff` on (a) the combined transcripts produced by Cuffcompare and (b) TopHat's accepted hits data sets for each data set. Cuffdiff produces quite a few output data sets; You will want to browse the Cuffdiff documentation to get a sense of what they do. Look at the isoform expression data set—are there any significant isoform expression differences between the two samples? Look at the isoform FPKM tracking data set—find an entry for a novel isoform and an entry for an isoform that matches a reference isoform. What is the nearest gene and transcription start site for each entry? (Hint: You will need to understand the class codes, which are explained in the Cuffcompare documentation.)

On Your Own

1. Identify all novel splice junctions and transcript isoforms in each set of transcripts.
2. Rerun Step 3, but assemble transcripts using the UCSC RefSeq genes as a reference. Find differences between de novo transcript assembly/analysis and reference-guided transcript assembly/analysis.

REFERENCES

Adams MD, Kelley JM, Gocayne JD, Dubnick M, Polymeropoulos MH, Xiao H, Merril CR, Wu A, Olde B, Moreno RF, et al. 1991. Complementary DNA sequencing: Expressed sequence tags and Human Genome Project. *Science* **252:** 1651–1656.

Adams MD, Dubnick M, Kerlavage AR, Moreno R, Kelley JM, Utterback TR, Nagle JW, Fields C, Venter JC. 1992. Sequence identification of 2,375 human brain genes. *Nature* **355:** 632–634.

Baltimore D. 1970. RNA-dependent DNA polymerase in virions of RNA tumour viruses. *Nature* **226:** 1209–1211.

Bank A, Terada M, Metafora S, Dow L, Marks PA. 1972. In vitro synthesis of DNA components of human genes for globins. *Nat New Biol* **235:** 167–169.

Bullard JH, Purdom E, Hansen KD, Dudoit S. 2010. Evaluation of statistical methods for normalization and differential expression in mRNA-Seq experiments. *BMC Bioinformatics* **11:** 94. doi: 10.1186/1471-2105-11-94.

Goecks J, Taylor J. 2011. Galaxy RNA-seq analysis exercise. http://main.g2.bx.psu.edu/u/jeremy/p/galaxy-rna-seq-analysis-exercise (accessed 8/10/2011).

Homer N, Merriman B, Nelson SF. 2009. BFAST: An alignment tool for large scale genome resequencing. *PLoS ONE* **4:** e7767. doi: 10.1371/journal.pone.0007767.

Langmead B, Trapnell C, Pop M, Salzberg SL. 2009. Ultrafast and memory-efficient alignment of short DNA sequences to the human genome. *Genome Biol* **10:** R25. doi: 10.1186/gb-2009-10-3-r25.

Li H, Durbin R. 2009. Fast and accurate short read alignment with Burrows–Wheeler transform. *Bioinformatics* **25:** 1754–1760.

Li M, Wang IX, Li Y, Bruzel A, Richards AL, Toung JM, Cheung VG. 2011. Widespread RNA and DNA sequence differences in the human transcriptome. *Science* **333:** 53–58.

Marioni JC, Mason CE, Mane SM, Stephens M, Gilad Y. 2008. RNA-seq: An assessment of technical reproducibility and comparison with gene expression arrays. *Genome Res* **18:** 1509–1517.

Mortazavi A, Williams BA, McCue K, Schaeffer L, Wold B. 2008. Mapping and quantifying mammalian transcriptomes by RNA-Seq. *Nat Methods* **5:** 621–628.

Pickrell JK, Marioni JC, Pai AA, Degner JF, Engelhardt BE, Nkadori E, Veyrieras JB, Stephens M, Gilad Y, Pritchard JK. 2010. Understanding mechanisms underlying human gene expression variation with RNA sequencing. *Nature* **464:** 768–772.

Rumble SM, Lacroute P, Dalca AV, Fiume M, Sidow A, Brudno M. 2009. SHRiMP: Accurate mapping of short color-space reads. *PLoS Comput Biol* **5:** e1000386. doi: 10.1371/journal.pcbi.1000386.

Temin HM, Mizutani S. 1970. RNA-dependent DNA polymerase in virions of Rous sarcoma virus. *Nature* **226:** 1211–1213.

Thierry-Mieg D, Thierry-Mieg J. 2006. AceView: A comprehensive cDNA-supported gene and transcripts annotation. *Genome Biol* (Suppl 1) **7:** S12.1–S12.14.

Trapnell C, Pachter L, Salzberg SL. 2009. TopHat: Discovering splice junctions with RNA-seq. *Bioinformatics* **25:** 1105–1111.

Trapnell C, Williams BA, Pertea G, Mortazavi A, Kwan G, van Baren MJ, Salzberg SL, Wold BJ, Pachter L. 2010. Transcript assembly and quantification by RNA-seq reveals unannotated transcripts and isoform switching during cell differentiation. *Nat Biotechnol* **28:** 511–515.

Verma IM, Temple GF, Fan H, Baltimore D. 1972. In vitro synthesis of DNA complementary to rabbit reticulocyte 10S RNA. *Nat New Biol* **235:** 163–167.

Wang ET, Sandberg R, Luo S, Khrebtukova I, Zhang L, Mayr C, Kingsmore SF, Schroth GP, Burge CB. 2008. Alternative isoform regulation in human tissue transcriptomes. *Nature* **456:** 470–476.

Zhulidov PA, Bogdanova EA, Shcheglov AS, Vagner LL, Khaspekov GL, Kozhemyako VB, Matz MV, Meleshkevitch E, Moroz LL, Lukyanov SA, Shagin DA. 2004. Simple cDNA normalization using Kamchatka crab duplex-specific nuclease. *Nucleic Acids Res* **32:** e37. doi: 10.1093/nar/gnh031.

WWW RESOURCES

http://main.g2.bx.psu.edu/u/jeremy/p/galaxy-rna-seq-analysis-exercise Galaxy RNA-seq Analysis Exercise.

http://wiki.g2.bx.psu.edu/Admin/Cloud The Galaxy Cloud project.

http://www.illumina.com/documents/seminars/presentations/2010-06_sq_03_lakdawalla_transcriptome_sequencing.pdf Illumina's preliminary data using DSN normalization for RNA-seq.

http://www.ncbi.nlm.nih.gov/IEB/Research/Acembly/index.html?human AceView genes, National Center for Biotechnology Information.

11

Metagenomics

Alexander Alekseyenko and Stuart M. Brown

Next-generation sequencing (NGS) technology has inspired a surge of research activity in microbiology. In particular, large-scale sequencing can provide detailed information regarding both the composition and the metabolic capabilities of communities of bacteria. This can be applied to environmental microbiology to study bacteria present in seawater (Venter et al. 2004), soil, or extreme habitats such as acid mine waste, or to study bacteria associated with the human body. It has been estimated that the human body hosts 10^{14} bacteria, 10 times more than the number of human cells. These bacteria are highly diverse. For example, the typical human gut may host 1000 different bacterial species. The total number of genes in these bacteria (the "metagenome") is perhaps 100 times greater than the number of genes in the human genome, giving them great metabolic diversity.

Human-associated bacteria are of particular interest to the U.S. National Institutes of Health (NIH) because they may be disease-causing pathogens or function as biomarkers for disease. The NIH initiated the **Human Microbiome Project** (HMP) in 2007 specifically to use NGS technologies to learn more regarding microbial communities associated with the human body (Turnbaugh et al. 2007). The HMP addresses four main goals:

- Determining whether individuals share a core human microbiome
- Understanding whether changes in the human microbiome can be correlated with changes in human health
- Developing the new technological and bioinformatics tools needed to support these goals
- Addressing the ethical, legal, and social implications raised by human microbiome research

These broad goals of the HMP led to a large set of specific experimental aims that include (1) surveys of microbes that inhabit specific body sites (Fig. 1); (2) functional

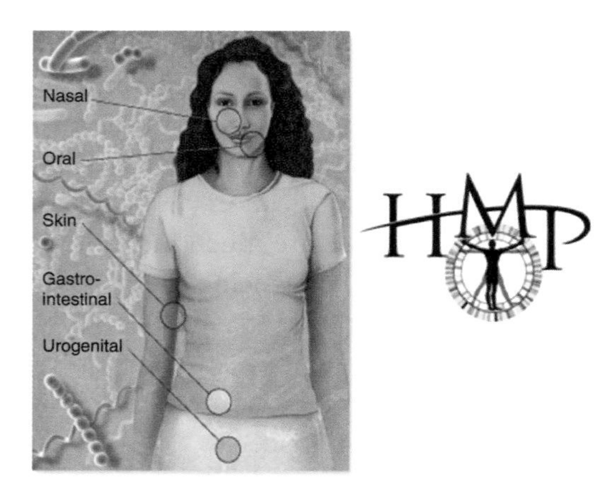

FIGURE 1. The Human Microbiome Project (http://commonfund.nih.gov/hmp/index.aspx).

studies of the genome content and the metabolic capabilities of bacteria at body sites; (3) assessment of the microbial contribution to human disease; and (4) determination of the mechanisms of interaction between complex microbial communities and the human host. Some examples of microbiome projects currently underway (in 2012) include the following:

- the role of antibiotics in obesity (Cho et al. 2012)
- microbiomic determinants of psoriasis (Blaser et al. 2010)
- microbial markers of foregut adenocarcinoma (Yang et al. 2010)
- microbial markers of severe early childhood caries
- the role of the microbiome in estrogen-driven malignancy
- the role of the gut microbiome in newborn immune development

NGS TECHNOLOGIES

Before the development of NGS, studies of bacterial communities were limited by sampling and identification methods. Classic microbiology relies on culture, dilution plating, and a combination of microscopy and staining to identify and count microbes in a sample. These methods are inherently limited to the study of microbes that grow well under specific sets of culture conditions. The development of PCR-based methods allowed for the discovery and quantitation of uncultured microbes, but the **cloning** and **Sanger sequencing** of individual gene fragments was laborious and expensive. NGS sequencing of PCR **amplicons** allows for much deeper sampling of microbial populations quickly and at low cost. This PCR-based method allows for the study of unculturable microorganisms. NGS **metagenomic** studies allow for the analysis of large numbers of samples, which allows investigators to address questions such as microbial diversity among a population of people or changes over time.

Most current metagenomic studies follow one of two basic methods, either to survey and count microbes using amplicon sequencing of a single gene (usually the 16S rDNA gene) and taxonomic informatics methods, or **shotgun** metagenomic sequencing and a collection of ad hoc informatics methods that include **de novo assembly**, gene identification, and species identification. As of 2012, the amplicon sequencing methods are much more advanced and in wider use.

The amplicon sequencing methods used in the HMP rely on "universal" PCR primers that target the **16S ribosomal DNA** gene. This gene is an essential component of the ribosome and is present in the genome of every known bacteria. The sequence of the 16S gene is composed of highly conserved regions, which are suitable for the design of multispecies PCR primers, and variable regions, whose sequence can be used to distinguish different bacteria at a meaningful taxonomic level (see Fig. 2). The HMP has chosen sets of 16S primers that produce amplicons of ~300–500 bp. These amplified fragments are typically sequenced using the **Roche 454 Genome Sequencer**, which produces reads of ~400 bp. These longer reads are necessary to acquire sufficient taxonomic information to identify bacterial species. Shorter reads cannot be joined by assembly because they may derive from templates isolated from different species. The 454 machine is capable of producing ~1 million reads per run, but this yield is typically subdivided by using barcodes in the sequencing adapters attached to each fragment, so that a single run of the sequencer can provide data for dozens of samples. Initial benchmarking studies found that the discovery of new species plateaus at about 10,000 fragments, so that is the targeted yield per sample. Note that it is not always possible to assign the sequence of a given PCR-amplified 16S fragment to a single bacterial species, but it can be assigned to a group at a higher taxonomic level. It is also true that counts of 16S sequences are not equivalent to counts of bacterial cells. Some bacteria have multiple copies of the 16S gene (up to 15 copies have been observed in a single bacterial genome), and these multiple copies may have identical or different sequences. The so-called universal 16S PCR primers create bias in the amplified sequences, so that the abundances of species (or other taxonomic units) observed with one set of primers are not comparable to those observed with other primers.

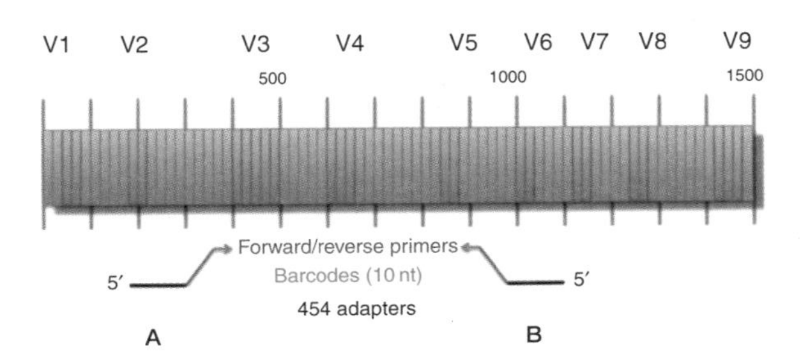

FIGURE 2. Location of the hypervariable regions (V1–V9) of the 16S rRNA and a typical primer construct for amplicon library preparation.

DATA ANALYSIS

Data analysis for 16S metagenomic studies involves cleaning and filtering sequence data, taxonomic analysis of 16S sequences, and clustering of samples (Wooley et al. 2010). Base calling for the 454 sequencer has characteristic error patterns, such as problems with insertion/deletion errors in homopolymer regions. Very short sequences (<200 bp) tend to be artifacts and therefore are usually discarded. The 3′ ends of sequences tend to have a higher error rate, thus a standard 3′ trim is often applied to all **sequence reads**. Because the 454 machine assigns a quality score to each base of each read, low-quality regions can be masked. The PCR process can create some artifacts such as chimeras (in vitro recombination between two different template molecules), which can be identified and removed with software such as ChimeraSlayer (Haas et al. 2011), UCHIME (Edgar et al. 2011), or Perseus (http://code.google.com/p/ampliconnoise) (Quince et al. 2011).

Taxonomic analysis of PCR-amplified 16S sequences can be performed by matching each sequence to its closest homolog in a database of **reference sequences**, or by **multiple alignment** and phylogenetic clustering. The most widely used reference databases for 16S sequences are the Ribosomal Database Project (RDP; http://www.cme.msu.edu) (Cole et al. 2009), greengenes at Lawrence Berkeley Laboratory (http://greengenes.lbl.gov) (DeSantis et al. 2006), and the microbial genomes section of **GenBank** at the NCBI (http://www.ncbi.nlm.nih.gov/genomes.lproks.cgi?view=1). The HMP has its own custom database of 16S data at the Data Analysis and Coordination Center (http://hmpdacc.org). It is not always possible to assign each 16S sequence to a unique species. Rather than provide a match with low statistical significance or an ambiguous match to several different database entries, the HMP uses taxonomic methods that assign the best match at the genus or a higher taxonomic level.

Without using a reference database, a set of 16S sequences (or any other multispecies amplicon) can be aligned to each other and clustered into operational taxonomic units (OTUs). These OTUs can be defined based strictly on specific amounts of sequence similarity (i.e., 3% sequence difference) or based on phylogenetic properties (neighbor joining, maximum likelihood, or Bayesian evolutionary models). Software for phylogenetic clustering includes PHYLIP (http://evolution.genetics.washington.edu/phylip.html) (Felsenstein 1989), FastTree (http://microbesonline.org/fasttree) (Price et al. 2010), BEAST (http://beast.bio.ed.ac.uk) (Drummond et al. 2012), and MrBayes (http://mrbayes.csit.fsu.edu) (Ronquist et al. 2012). Then samples can be clustered based on numbers of shared OTUs. Samples can also be interrogated for species abundance, richness, and diversity. A simple hypergeometric test can be used to identify over-representation of individual phyla, which may be important, even if no significant differences are found when samples are clustered based on all sequences or all OTUs. When performing statistical tests for individual phyla or OTUs, it is important to control for multiple testing.

Another approach to clustering samples is used by the program UniFrac (Lozupone et al. 2011). UniFrac builds a single phylogenetic tree from all sequences from all samples in a study. Distances between pairs of samples are calculated across branches of the tree that contain OTUs that are shared and unshared by a pair of samples by the formula: (sum of unshared branch lengths)/(sum of shared + unshared). This is equivalent to saying that the distance between two samples is the fraction of branch lengths that are unshared (see Fig. 3).

Rather than gathering an ad hoc collection of different software, it is helpful to use an integrated package of tools that have been built to support **metagenomics** studies. The QIIME package (Quantitative Insights Into Microbial Ecology) is the most widely used (http://qiime.sourceforge.net). QIIME includes PyNAST alignment, tree building, taxonomy assignment, OTU picking, and clustering and visualization by Principal Coordinate Analysis (PCoA). QIIME also includes scripts that automate many of the standard analysis pipelines used in metagenomic analysis for the HMP and other projects. QIIME is distributed as a virtual machine, which can be installed on top of the VirtualBox, a free open source virtual operating system that can be run on Windows, Macintosh, Linux, and Solaris computers. By installing QIIME with VirtualBox, users can be guaranteed a consistent workspace environment with minimal hassles from software installation or incompatibility. mothur is another project that has developed a single piece of open source, expandable software to fill the bioinformatics needs of the microbial ecology community (Schloss et al. 2009). It includes the functionality of DOTUR, SONS, TreeClimber, s-libshuff, and UniFrac, as well as calculators and visualization tools. mothur is distributed as Macintosh, Windows, and Linux executables as well as open source code (http://www.mothur.org/wiki/Main_Page).

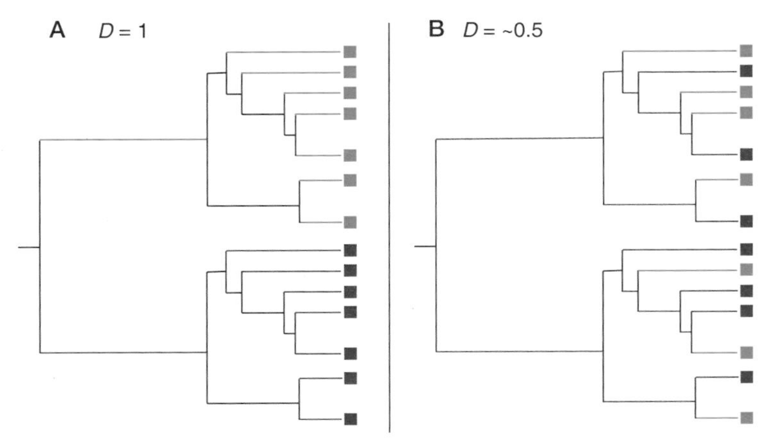

FIGURE 3. Distances between 16S sequence reads derived from two different samples (red and blue squares) measured as unique branch lengths in a phylogram. (*A*) Two samples have no common or closely related species, UniFrac distance = 1. (*B*) Two samples have related species, but unique branch lengths are observed, UniFrac distance = 0.5.

TUTORIAL

A tutorial is available online (http://qiime.org/tutorials/tutorial.html) including an install of VirtualBox and QIIME and standard analysis of a 454 HMP data set (Crawford et al. 2009) using standard QIIME tools and scripts.

- Create a mapping file, which is a collection of project metadata.
- Separate the sequences by barcode.
- Quality analysis of reads and filtering of low-quality data.
- Cluster sequences into OTUs and assign OTUs to taxonomic identity.
- Descriptive analysis (abundance, diversity, richness).
- PCoA.

REFERENCES

Blaser MJ, Methe B, Strober B, Perez Perez GI, Brown S, Alekseyenko A. 2010. Evaluation of the cutaneous microbiome in psoriasis. *Nature Precedings* doi: 10.1038/npre.2010.5276.1.

Cho I, Yamanishi S, Cox L, Methé BA, Zavadil J, Li K, Gao Z, Raju K, Teitler I, Li H, et al. 2012. Antiobiotics in early life alter the murine colonic microbiome and adiposity. *Nature* **488:** 621–626.

Cole JR, Wang Q, Cardenas E, Fish J, Chai B, Farris RJ, Kulam-Syed-Mohideen AS, McGarrell DM, Marsh T, Garrity GM, Tiedje JM. 2009. The Ribosomal Database Project: Improved alignments and new tools for rRNA analysis. *Nucleic Acids Res* **37:** D141–D145.

Crawford PA, Crowley JR, Sambandam N, Muegge BD, Costello EK, Hamady M, Knight R, Gordon JI. 2009. Regulation of myocardial ketone body metabolism by the gut microbiota during nutrient deprivation. *Proc Natl Acad Sci* **106:** 11276–11281.

DeSantis TZ, Hugenholtz P, Larsen N, Rojas M, Brodie EL, Keller K, Huber T, Dalevi D, Hu P, Andersen GL. 2006. greengenes, a chimera-checked 16S rRNA Gene Database and workbench compatible with ARB. *Appl Environ Microbiol* **72:** 5069–5072.

Drummond AJ, Suchard MA, Xie D, Rambaut A. 2012. Bayesian phylogenetics with BEAUti and the BEAST 1.7. *Mol Biol Evol* **29:** 1969–1973.

Edgar RC, Haas BJ, Clemente JC, Quince C, Knight R. 2011. UCHIME improves sensitivity and speed of chimera detection. *Bioinformatics* **27:** 2194–2200.

Felsenstein J. 1989. PHYLIP—Phylogeny Inference Package (Version 3.2). *Cladistics* **5:** 164–166.

Haas BJ, Gevers D, Earl AM, Feldgarden M, Ward DV, Giannoukos G, Ciulla D, Tabbaa D, Highlander SK, Sodergren E, et al. 2011. Chimeric 16S rRNA sequence formation and detection in Sanger and 454-pyrosequenced PCR amplicons. *Genome Res* **21:** 494–504.

Lozupone C, Lladser ME, Knights D, Stombaugh J, Knight R. 2011. UniFrac: An effective distance metric for microbial community comparison. *ISME J* **5:** 169–172.

Price MN, Dehal PS, Arkin AP. 2010. FastTree 2—Approximately maximum-likelihood trees for large alignments. *PLoS ONE* **5:** e9490. doi: 10.1371/journal.pone.0009490.

Quince C, Lanzen A, Davenport RJ, Turnbaugh PJ. 2011. Removing noise from pyrosequenced amplicons. *BMC Bioinformatics* **12:** 38. doi: 10.1186/1471-2105-12-38.

Ronquist F, Teslenko M, van der Mark P, Ayres DL, Darling A, Höhna S, Larget B, Liu L, Suchard MA, Huelsenbeck JP. 2012. MrBayes 3.2: Efficient Bayesian phylogenetic inference and model choice across a large model space. *Syst Biol* **61:** 539–542.

Schloss PD, Westcott SL, Ryabin T, Hall JR, Hartmann M, Hollister EB, Lesniewski RA, Oakley BB, Parks DH, Robinson CJ, et al. 2009. Introducing mothur: Open-source, platform-independent, community-supported software for describing and comparing microbial communities. *Appl Environ Microbiol* **75:** 7537–7541.

Turnbaugh PJ, Ley RE, Hamady M, Fraser-Liggett CM, Knight R, Gordon JI. 2007. The Human Microbiome Project. *Nature* **449:** 804–810.

Venter JC, Remington K, Heidelberg JF, Smith HO, Rusch D, Eisen JA, Wu D, Paulsen I, Nelson KE, Nelson W, et al. 2004. Environmental genome shotgun sequencing of the Sargasso Sea. *Science* **304:** 66–74.

Wooley JC, Godzik A, Friedberg I. 2010. A primer on metagenomics. *PLoS Comput Biol* **6:** e1000667. doi: 10.1371/journal.pcbi.1000667.

Yang L, Oberdorf WE, Gerz E, Parsons T, Shah P, Bedi S, Nossa CW, Brown SM, Chen Y, Liu M, et al. 2010. Foregut microtome in development of esophageal adenocarcinoma. *Nature Precedings* doi: 10.1038/npre.2010.5026.1.

WWW RESOURCES

http://beast.bio.ed.ac.uk BEAST.

http://code.google.com/p/ampliconnoise Perseus.

http://commonfund.nih.gov/hmp/index.aspx Human Microbiome Project homepage.

http://evolution.genetics.washington.edu/phylip.html PHYLIP.

http://greengenes.lbl.gov greengenes at Lawrence Berkeley Laboratory.

http://hmpdacc.org HMP's custom database of 16S data at the Data Analysis and Coordination Center.

http://microbesonline.org/fasttree FastTree.

http://mrbayes.csit.fsu.edu MrBayes.

http://qiime.org/tutorials/tutorial.html QIIME tutorial.

http://qiime.sourceforge.net QIIME (Quantitative Insights Into Microbial Ecology) package.

http://www.cme.msu.edu Ribosomal Database Project (RDP).

http://www.mothur.org/wiki/Main_Page mothur.

http://www.ncbi.nlm.nih.gov/genomes.lproks.cgi?view=1 Microbial Genomes section, GenBank, NCBI.

12

High-Performance Computing in DNA Sequencing Informatics

Efstratios Efstathiadis and Eric R. Peskin

Since the turn of the century, we have witnessed a synthesis of information technology and science, frequently labeled as *eScience*, fueled in part by the enormous amounts of digitized data produced by high-throughput experimental devices (genome sequencers, mass spectrometers, digital scanners, etc.). Currently, **next-generation sequencing** (NGS) instruments are capable of producing billions of DNA **sequence reads** in a single run, which translates to several hundred gigabytes of raw data. As the number and size of NGS data sets increase rapidly, the diversity, complexity, and distribution of data and their resources also escalate. Preparing, managing, and analyzing diverse data sets that come in a variety of types and formats and from heterogeneous sources make data integration and discovery a challenging task that is beyond the computing skills of many scientists. To execute large-scale analysis and support effective data mining, a tight integration of information is needed requiring the orchestration of many tools.

WORKFLOWS

Workflows (Romano 2008; Deelman et al. 2009; Goble and De Roure 2009) provide a systematic, automated means of accessing and integrating data from a pool of resources as well as conducting analysis across diverse data sets and applications. The main goal of a workflow is the implementation of data analysis processes in standardized environments. The advantages of workflows relate to the following:

- *Effectiveness*. Being an automatic procedure, workflows can liberate bioscientists from routine and repetitive tasks, such as capturing, moving, and archiving data. By hiding low-level details, workflows empower a wide spectrum of scientists to access data resources and execute sophisticated pipelines.

- *Best Practice.* Workflows can be designed and validated by experts, then shared with an unlimited number of other users who wish to use the same analytic methods.
- *Reproducibility.* Over time, analysis can be replicated and validated.
- *Reusability.* Intermediate results of processes can be reused and adapted.
- *Traceability.* A transparent analysis environment is present for workflow execution.

In creating and executing NGS data analysis pipelines, workflows describe a multistep process, coordinate multiple tasks, and provide the flexibility that is needed to cope with the diversity of data. Each task in the complex analysis process represents the execution of a computational process (such as running a program), which can be ordered and interlinked with other tasks so that the process can be performed by executing each task when all of the needed requisites are fulfilled. Examples of such tasks can be data transfer, data compression, executing database queries, mapping sequencer reads to a **reference genome**, converting between file formats, and submitting a job to a resource manager. The output of one task is consumed by subsequent tasks according to a predefined schema that "orchestrates" the transient data flow between data sources and analytical tools.

Workflows can simplify the analysis and interpretation of massive data sets generated by NGS instruments using widely applied and the best-validated analysis methods currently available. NGS workflows must be adaptable to novel applications, new **algorithms**, and computational methods that are rapidly emerging in the NGS landscape. Highly customizable workflows can enable researchers to produce their own, custom analysis workflows, modify and integrate existing ones, and share them with collaborators. Workflows integrated with web services can grant to scientists access to sophisticated applications without the need to install and operate them and thus use the best applications, not just the ones they are familiar with.

Taverna (http://www.taverna.org.uk/) (Hull et al. 2006) is a scalable, open source workflow management system used for designing and executing workflows. It orchestrates a wide range of services to support workflows for analyzing genomic data. It uses a powerful and flexible language to describe complex workflows, such as iterations and other well-understood programming language constructs. Taverna provides a generic external plug-in framework that allows a range of programming frameworks to be incorporated including web services, grid services, databases, command-line tools, Java applications and libraries (like the Chemistry Development Kit), and a Portable Batch System (PBS) plug-in that allows the execution of workflows on an HPC cluster. Using a graphical user interface, users can create workflows composed of interconnected processes. Although Taverna workflows are created by computational scientists, the resulting workflows are provided as "wrapped-up" processes on portals and are executed on a Taverna Server, allowing scientists with limited computing background, technical resources, and support to construct highly complex analyses over public and private data. The Taverna Workbench, a free, easily installed desktop application available on most popular computer platforms provides a user-friendly environment to start creating, editing, and running workflows.

Taverna has integrated interfaces to *Bio*Catalogue (http://www.biocatalogue.org/) (Bhagat et al. 2010) and myExperiment (http://www.myexperiment.org/) (Goble et al. 2010). *Bio*Catalogue provides a curated catalog of life sciences web services where services can be registered by providers, discovered by scientists, annotated with descriptions and tags, and monitored to check availability and functionality of volatile services. myExperiment is a workflow repository where workflow developers can privately store or allow open public access to their workflows. myExperiment can act as a social website that enables social networking around workflows. It gathers comments, ratings, recommendations, and mixing of new workflows with those previously deposited. Alternatively, organizations can take advantage of myExperiment features and strengths by establishing their own local myExperiment server for internal sharing of workflows. Taverna plug-ins can search for myExperiment workflows and import them in a developing workflow.

Web portals provide researchers with access to a number of predefined workflows and enable the enactment of workflows in a user-friendly environment, thereby freeing researchers from the burden of developing workflows. Additionally, portals allow users to store execution metadata and related results. Galaxy (http://galaxyproject.org) is a popular, open source, web-based genomic workbench that enables users to perform integrative computational analysis of genomic data. It provides an interactive web interface (http://usegalaxy.org/) for obtaining genomic data and a unified method for accessing and applying computational tools when analyzing data (Blankenberg et al. 2010; Goecks et al. 2010). Users can import data sets into their workspaces from many established warehouses or upload their own data sets. Galaxy supports the complete life cycle of a computational biology experiment by providing the following.

- Easy access to a large set of computational tools for researchers via a web interface through which tools can be included in an analysis chain and run by scientists without programming experience. Custom tools (written in any language) can be integrated with the rest of the tools in an analysis chain, provided that a web tool or a command-line invocation of the tool can be constructed.
- Automatic capture of metadata, including every piece of necessary information to ensure repeatability of each step, when a user performs an analysis using Galaxy (data, program, parameters). Along with automatically provided metadata, user-provided metadata, such as annotations (notes regarding each analysis

step) and tagging (labeling), are made available in the history panel. Users can create, copy, and version histories, making entire analysis chains reproducible.

- An analysis history serves as an easy method to create workflows that can run repeatedly (the same tools with the same parameters) on different data sets. Alternatively, using the graphical interface of the workflow editor, users can create workflows from scratch. Galaxy facilitates transparency by enabling users to share and communicate their experimental results in a meaningful way (providing access to shared data sets, histories, and workflows and by enabling users to communicate their experiments at every level of detail) so that others can view, reproduce, and extend their experiments.

In addition to the public server (http://usegalaxy.org/), Galaxy is distributed as a stand-alone package and can be deployed and configured on an external web server that is local to researchers' sites (a Galaxy Instance), a very useful feature when large data sets need to be uploaded to be included in analysis chains. Galaxy can be interfaced with any DRMAA-compliant (http://drmaa.org/) resource management system such as Sun Grid Engine (SGE) and Portable Batch System (PBS) to use compute clusters for running jobs. At the 2011 Galaxy Community Conference, Illumina presented the first test results (gap analysis) of running their proprietary secondary analysis software (CASAVA) workflows using Galaxy. Galaxy can be deployed within an institution as a user-friendly gateway to local repositories of data, software, computing power, and validated best-practice workflows.

Taverna workflows can be incorporated in Galaxy as custom tools, combining the features of both frameworks: the easy, unified access to bioinformatics tools that Galaxy offers and Taverna's powerful workflow expressions (Karasavvas et al. 2011). Given a Taverna workflow and having access to a Taverna server, a Galaxy tool generator will generate the XML configuration file and the processing file that can be deployed (currently manual installation of the tool is required) on a Galaxy server. All Taverna2 public workflows available on myExperiment provide access to the Galaxy tool generator by a simple click on the "Download Workflow as a Galaxy tool" option on the workflow description web page. The command line on a Taverna2 workflow file (.t2flow) can also invoke the generator to create a zip file that includes the Galaxy tool configuration and processing files.

Galaxy Cloudman enables the instantiation, running, and scaling of cloud resources with an automatically configured Galaxy. Galaxy and the Galaxy-required tools have been packed as a Virtual Machine (VM) image that resides on Amazon Web Services (AWS). The VM is easily instantiated, offering the same functionality as any other instance of Galaxy. The latest Galaxy Amazon Machine Images (AMI) names and IDs are available at http://usegalaxy.org/cloud.

The NSF-funded XSEDE (Extreme Science and Engineering Discovery Environment; https://www.xsede.org/web/guest/data-transfers) project is a follow-on to the

TeraGrid project that supports several supercomputers and high-end visualization and data analysis resources across the United States. Resources are provided by several partner institutions known as service providers (SPs). The project integrates advanced digital resources and services, makes them easier to use, and helps increase the number of researchers who use the resources. XSEDE is a partnership of a number of institutions across the United States that is supported by the National Science Foundation.

SOFTWARE TOOLKITS

Performing complex analysis on locally available data sets (either generated by local instruments or downloaded from elsewhere) may require application-specific workflows, involving direct invocation of locally installed tools, giving users the flexibility to tweak several input parameters. Several open source, freely available, community-developed toolkits, such as BioPerl (Stajich et al. 2002), Biopython (Cock et al. 2009), BioJava (Holland et al. 2008), and Bioconductor (Gentleman et al. 2004), have been created on top of fully featured programming or scripting languages, which enable application development for a wide range of sequencing projects. Such tools require some level of programming and scripting expertise to fully use their potential and are often not suitable for the average biologist, although significant effort is being made to hide technical details from the users.

BioPerl (http://www.bioperl.org) is a toolkit consisting of a collection of reusable Perl modules (self-contained pieces of code and associated data) that enables users to develop bioinformatics workflows. Users can write their cross-platform scripts, reducing powerful bioinformatics tasks to only a few lines of code, taking advantage of Perl's strengths. Perl's regular expressions are used to parse **GenBank** and **BLAST** reports. Database connectivity modules are used in submitting queries and retrieving data from archives. Perl's flexibility in managing large amounts of data assists in processing the output of high-throughput sequencers. BioPerl is an object-oriented toolkit in which multiple modules depend on each other to achieve a task. Modules group together to form packages such as the core package, which is required by all other packages, and the run package, which contains wrappers for executing common bioinformatics applications. CPAN (Comprehensive Perl Archive Network; http://www.cpan.org) provides a huge collection of well-organized, freely available, tested modules and scripts, as well as online documentation and tutorials.

Bioconductor (http://www.bioconductor.org/) is an open source project that provides a large set of R packages (http://www.r-project.org/) for statistical data analysis in the area of life science, such as **microarray** and genome analysis. Bioconductor relies on the strengths of the R programming environment that provide access to facilities for advanced statistical analysis, access to high-quality numerical routines, data visualization, and transformation tools, in order to support advanced operations in sequence-based analysis such as integration with diverse genomic resources and transformation among sequence-related file types. A large number of Bioconductor packages are freely available for high-throughput sequencing tasks, such as *ShortRead* (Morgan et al. 2009), used to support common and advanced sequence manipulation operations (trimming, transformation, and **alignment**), *Biostring* used for alignment and pattern matching, an interface to SAMtools (*Rsamtools*), interfaces to Sequence Read Archives (*SRAdb* package), and facilities for **ChIP-seq** and related activities. Sample workflows for processing sequence data and a list of resources may be found at http://www.bioconductor.org/help/workflows/high-throughput-sequencing/. Similarly to Perl, the Comprehensive R Archive Network (CRAN) (http://cran.at.r-project.org) provides a large collection of code and documentation mirrored on a network of web and FTP servers. In addition, an Amazon Machine Image (AMI) has been developed and optimized for running Bioconductor in the Amazon Elastic Compute Cloud (EC2) for sequencing tasks, preloaded with R and several Bioconductor (and CRAN) packages (http://www.bioconductor.org/help/bioconductor-cloud-ami/). Whereas BioPerl is more suitable for processing large amounts of sequencing data and interfacing with sequencing databases, Bioconductor is suitable for statistical analysis but is not well suited for large data sets and lacks a bytecode compiler.

VERSION CONTROL

When developing custom solutions, revision control systems prove to be great collaborative tools because they make it easy to share and co-develop software. Revision or version control systems or SCM (for Source Code Management) refer to the tools that manage changes and track the history of revisions in programs, scripts, documents, and all other files that make up a software package. Central to the concept of revision control is the *repository* (a tree of files), where all important project files are gathered.

There are several free SCM systems available on most popular platforms, such as git, Mercurial, Bazzar, subversion, and CVS (Concurrent Versions System). The first generation of SCMs is either locally accessible on a single server (RCS) or provides a centralized repository that hosts several projects and is accessible over the network in a client-server architecture. Authorized users may access the centralized repository to check out packages and upload their changes. Most recent version control systems are distributed (decentralized), where there is no "special" central repository. Developers copy the entire repository, including all project modifications and development history of the project on their local servers. Project modifications are imported as branches into the developer's repository, where they can merge with locally developed branches. Distributed systems are beneficial because development does not depend on network connectivity as changes are committed to a local repository rather than remote ones. In addition, they are unaffected by slow network connections or overloaded centralized repository servers and are better load-balanced, because changes from developers are committed to separate servers. Moreover, they eliminate the single points of failure that centralized repositories suffer from.

git (http://git-scm.com/) is a fast, robust, open source version control system, optimized for distributed development of small or large projects. It was initially created as a replacement of BitKeeper for the development of the Linux kernel. As a distributed version control system, git does not depend on a central server or network connectivity or downtime. Thus, each developer does not need a special account to be able to commit changes in central repositories as in CVS or subversion. Revision control systems are not just for large projects where a large number of developers are involved and thousands of code lines are being written. They are straightforward to understand, simple to use, easy to install, and come with very good documentation that makes it trivial to get new users started. Even small workflow development efforts can benefit greatly from such tools.

DATA STORAGE

A series of technological innovations has increased the efficiency of NGS instruments several orders of magnitude from previous generation sequencing techniques. Starting around 2005, when the first NGS instruments became available in the market, the sequencing output has doubled every few months, outpacing both computing performance, approximated by Moore's law, and hard disk drive storage capacity, approximated by Kryder's law (Kahn 2011). It is estimated that over the last four years the throughput of sequencing instruments has increased by a factor of 1000, while computing power has increased by approximately a factor of 4.

With current NGS technology, whole genome sequencing requires 30 or more times **coverage**. Such a level of coverage is necessary for several reasons: The uneven distribution of short individual reads will cover some regions of the genome multiple times, but other regions will not be covered at all; repetitive, hard-to-sequence regions require additional coverage; and multiple coverage helps identify machine errors. The quality score that is assigned to every base further increases the size of data files that must be stored. The cost of genome sequencing is now decreasing several times faster than the cost of data storage, promising that soon it will cost less to sequence a base of DNA than to store it on hard disk. Illumina's HiSeq 2000 sequencing systems, for example, can sequence 200 billion base pairs (200 Gb) in a week, a sequencing volume that is more than 60 times the coverage of the 3 billion base pairs in the human genome. Data output from today's single NGS run would have been considered an enormous amount of sequencing data in the beginning of the century, when it took several years to sequence the human genome using conventional **Sanger**

sequencing techniques. In parallel, the plummeting sequencing costs have enabled small laboratories and academic institutions around the globe to purchase and operate an increasing number of NGS instruments, producing massive amounts of sequencing data, making NGS one of the most data-intensive fields in the life sciences. Collected data sets must be moved off the instruments, stored on sophisticated data storage systems, analyzed by computing clusters supporting complex computational workflows, and archived, making data storage devices, along with HPC clusters, necessary components of every IT infrastructure that supports NGS.

Until recently, the raw output of NGS instruments was in the form of binary images, resulting in several hundred gigabytes of image data per run, requiring large data capacity to transfer and store the raw data. More recently, NGS instruments have integrated image analysis and base-calling functions, processing and deleting images in real time, and incorporating data reduction and compression in the upstream pipeline. The raw output of instruments, even those that use optical technologies, is now changing from large image files to intensity files, base calls, and quality scores. In addition, a new generation of NGS instruments is emerging, based on nonoptical technologies, which produce data in a variety of formats. As the technology of sequencing instruments is progressing, storage requirements for the downstream analysis are increasing. Although the instrument's expected output can be easily calculated based on the length of individual reads (number of base pairs), number and type of reads (single-end vs. **paired-end**), and type of study (**RNA-seq**, ChIP-seq, **de novo sequencing**, SNP discovery, etc.), the storage requirements for the secondary analysis are far more difficult to predict, considering the variety of data types with different characteristics and different data analysis software, as well as metadata and intermediate results.

Secondary (downstream) analysis includes several tasks that are both CPU and data intensive, such as mapping **short reads** to a reference genome, **assembly** process (de novo sequencing), removal of duplicate alignments, and so on. As an example, we consider the amount of sequencing data (reads) from a mouse genome collected and aligned in a paired-end run using eight flow-cell lanes on an Illumina HiSeq 2000 sequencer. A total of 14 FASTQ (Cock et al. 2010) files are generated using CASAVA (Consensus Assessment of Sequence and Variation) (Ilumina 2011a,b), each file containing 120 million (120×10^6) reads, each read being 51 bp long. In a paired-end run, we have two **FASTQ files** for each of the lanes, with each FASTQ file taking 20 GB of disk space. The total amount of sequencing in this example represents about five times the coverage of the reference (mouse) genome, which is considered a rather low level of coverage. The reads are then aligned to a reference mouse using SAMtools (Li et al. 2009) and Burrows–Wheeler Aligner (BWA) (Li and Durbin 2009). The Sequence Alignment/Map (SAM) text files that contain the read alignments generated using the BWA **sequence alignment** software are > 50 GB in size. The corresponding binary **BAM files** are ~15 GB and the sorted BAM files

10 GB. Two intermediate files are generated containing the suffix array coordinates (with .sai extension), each taking up 3.5 GB. In BWA, mapping of a short nucleotide read is equivalent to searching the *suffix array interval* for substrings of the chromosome that match the short read. Overall, the total amount of data generated for just the alignment part of this study is close to 1 TB, and it takes a couple of days to complete.

A typical sequencing laboratory includes shared data storage systems for the output of instruments as well as intermediate data and results generated by secondary analysis. A laboratory may run several sequencing instruments, from different vendors, with each instrument generating data with a variety of characteristics, formats, and volume. Multiple experiments may be running simultaneously writing large volumes of data to a networked storage system that is shared with computing resources, such as an HPC Linux cluster. The HPC cluster is a shared resource that performs most of the computing intensive data analysis in the form of parallel batch jobs. Having a storage system shared among various resources via NFS or CIFS eliminates the need for transferring and duplicating large data sets. A shared storage system enables individual researchers to monitor the status of a running experiment, access analysis results, or perform additional analysis and visualization on their local workstations. It provides data access to web-based analysis tools, workbenches, workflow execution tools, and visualization tools such as a local Galaxy deployment, Genome Browsers, and the like. Laboratory Information Management System (LIMS) is used in laboratories as a means to provide a single interface to track samples, tests, and results that may also require access to stored sequencing results. Clearly, the combined workload from the above functions is mixed and unpredictable. The challenge in data management for NGS is not merely to provide the necessary data storage capacity (a petabyte of data storage is readily available from several storage vendors), but rather to manage the changing variety and complexity of the data sets and access patterns. Research needs, however, change fast, as do instruments' output, application codes, and workflows; and changes happen at rates faster than the underlying IT infrastructure that supports storage.

There is no best-practice solution for selecting a data storage system for an NGS lab. The appropriate solution is both site and use specific. The needed data storage capacity and system type depend on the number of sequencing instruments and services that are being supported, how they are being used, and the type of projects in which the laboratory specializes. What helps is a good understanding of how much data are being produced, what are the data rates, and what type of access is required (type and number of clients, IO patterns, etc.). Small laboratories may decide to use a RAID-based Network Attached Storage (NAS) system as the primary data storage system. At the other end of the spectrum, laboratories with many sequencing instruments deploy highly scalable, multitiered, clustered file systems, with automatic data replication and backups. Here are some things to consider when selecting a data storage system to support the NGS storage infrastructure.

- A close collaboration between IT and researchers is needed to understand what data must be stored, archived, and backed up. In addition, it is important to understand what services (Taverna, Galaxy, LIMS, GBrowse, etc.) need to access the data and how these services are provided. Users need to be educated regarding the true costs of data storage and get involved in setting expectations. They need to understand the negative economy of scale in data storage: As the data storage capacity increases, the cost per terabyte also increases because additional mechanisms need to be implemented to account for data redundancy, protection, availability, multiple data access paths, archives, electrical power, cooling, rack space, dedicated network equipment, and so on. There is a disconnect between researchers (consumers of storage) and storage providers regarding what it costs to provide a large data storage system. A terabyte of data storage is not large by today's standards and can be purchased over the counter at any office supply store for less than $100. Small NGS laboratories that operate a single instrument, in an effort to economize, may accumulate portable USB Hard Disk Drives (HDDs) on laboratory benches. However, such unstructured systems cannot easily account for drive failures or online network access and are soon abandoned as data management becomes impossible.
- Cost and performance (read/write) are important factors in selecting a data storage system in every field. The selected system must provide the needed throughput and must be within the available budget. Stability and redundancy are also important factors, because some sequencers need reliable, uninterrupted access to the data storage system during a run.
- The data storage system must provide clients for the most popular operating systems (Linux, Windows, Mac OS) and access over the network to both, Windows servers via the Common Internet File System (CIFS), and Linux-base servers and Mac OSX via the Network File System (NFS). For example, Illumina GAIIx and HiSeq-2000 sequencing systems operate a control PC that runs Windows OS, requiring direct access to the data storage systems via CIFS. On the other hand, Linux clusters access shared file systems via NFS.
- If your budget allows, multitiered storage systems provide several benefits. Tier-1 provides the fastest, most reliable, and stable storage for the production needs. This is where the sequencer output will first be stored and the HPC cluster will analyze the data. Tier-2 consists of slower hard drive disks used for data sets that are not accessed as often. Tier-3 storage may even be provided by tapes. Most tiered systems automatically move files from one tier to another based on predetermined policies, such as last file accessed time.

- A data storage system must fit with the existing infrastructure. The bandwidth and traffic of the existing enterprise network, the network path from the sequencer to the storage, data center space, and electrical power, should be taken into account. Energy efficient solutions should be considered. Large storage systems may occupy several racks of space and use up kilowatts of power and require substantial AC for cooling. The cost for power and cooling adds to the total cost of ownership.
- Laboratories with small local support groups often rely on vendor support to install, configure, maintain, and troubleshoot data storage systems. Choosing a system from a vendor with known reliable support is a plus. Parallel file systems, for example, can perform poorly because of bad planning and configuration. Implementations of parallel file systems where small metadata files are stored in RAID5 or 6 arrays respond rather poorly when metadata files are accessed. It is always a good idea to consult with peer institutions regarding their experiences with the support of their storage vendor.
- The scalability of the storage system is important. Backing up large data storage systems is a real challenge. Part of the budget should be allocated for backups. As the price of sequencing drops, it may be more cost efficient and faster to repeat the experiment rather than store data or recover lost data.

A centralized data storage system simplifies, and in some cases eliminates, the need for transferring data locally within the laboratory or the institution. Data sets can be made available to a number of services without being moved or duplicated. In a UNIX environment, symbolic links (*symlinks*) can provide data access to several different services. Data sets can be shared between servers within the local network via CIFS and NFS without copying, which reduces network traffic and storage needs. Servers that do not have direct access to a shared storage system require data to be transferred over the local network. There is a plethora of tools that can facilitate the transfer of data securely, such as secure copy (*scp*), *rsync*, and *sftp*. Some of the tools also have GUI-enabled versions (*Winscp*, *Filezilla* for Windows, *Fugu* for Mac OS X). These tools can encrypt either the authentication step or the entire session, sending only encrypted information over the network. There are clients of the tools available on most common platforms. Data sets are frequently downloaded from a web server, either using a web browser or command-line tools, such as *curl* and *wget*.

Sharing large data sets with other researchers over the Internet at remote sites is more challenging, because of firewall issues, complex security issues, and unreliable networks. Sharing data is an important step in collaborative science, and it increases the transparency of the scientific process by allowing for scientific results to be independently tested and verified. NIH declared that the sharing of data is essential in translating research into knowledge and products that improve health. Several funding agencies have policies that support data sharing and encourage investigators to make their data available. Journals such as *Nature*, *Science*, and *Cancer Research* have policies that urge authors to deposit specific types of data on their relevant repository and submit accession numbers for deposited sequences with the manuscript. BioSharing (http://biosharing.org/) maintains a list of funding agencies' data-sharing policies. Although policies are in place and technologies to support them exist, the exact mechanisms for data sharing remain unspecified, and the implementation is untested (Savage and Vickers 2009).

The tools commonly used to transfer data over local networks have certain limitations that make them unsuitable for transferring data over wide area networks. The Energy Sciences Network has extensive information of the internal buffer limitation of TCP-based tools (http://fasterdata.es.net/fasterdata/say-no-to-scp/). GridFTP (Allcock 2005) is an extension of the standard File Transfer Protocol (FTP) for use with Grid Computing that addresses such limitations. It is part of the Globus Toolkit (http://www.globus.org/), a set of tools that enables the building of grids that allow distributed computing power, storage resources, scientific instruments, and other tools to be shared securely across corporate, institutional, and geographic boundaries. Key features of GridFTP include using multiple streams in parallel between a single source and destination (improving aggregate bandwidth relative to that achieved by a single stream), X509 certificates for authentication and encryption, third-party transfers, partial file transfers, fault tolerance and restart, automatic TPC optimization, and UDP-based file transfers (offering several fold speed improvements over TCP-based transfers). It is important, however, to make data transfers transparent and easy for researchers. Investigators generally do not have the IT expertise to develop, execute, monitor, and restart large file transfers. *GlobusOnline* (https://www.globusonline.org/) is an easy to use, grid-based tool, hosted on the Amazon Web Services cloud that can improve the speed of file transfers several times faster than *scp* and *sftp*.

As the volume of sequencing data keeps increasing at ever higher rates, so does the need for computing resources to analyze it. The data analysis and knowledge extraction step is often more time-consuming and more expensive than running the sequencing machine. As sequencing costs drop at rates faster than Moore's law, the relative cost of analysis versus sequencing is constantly increasing. **High-performance computing** (HPC) clusters can provide the needed computing power to run intensive analysis pipelines. Several NGS codes scale well and benefit from multithread processing. However, parallelizing codes requires significant effort, and Amdahl's law limits performance improvements. With the introduction of the Compute Unified Device Architecture (CUDA) from NVIDIA, graphics processing units (GPUs) have become more easily programmable, resulting in several bioinformatics applications being GPU-enabled, such as BarraCUDA (Klus et al. 2012) and SOAP3 (Liu et al. 2012).

High-performance computing clusters consist of a large number of computing nodes interconnected via a dedicated, private, high-speed network in order to enable the passing of messages and data of parallel tasks among processes. A common set of software packages are installed on all nodes, making a cluster partition behave like a single, unified resource. Most HPC clusters run a version of Linux. The head node, with access to the public/enterprise network, is where users log in interactively, edit, compile, debug and submit jobs, and transfer data, whereas the computing nodes are the workers that execute the programs and have multicore CPUs and several gigabytes of RAM. The configuration of the nodes (architecture, number of processing cores, memory size, and access speed) is a critical component that needs to be designed carefully based on application requirements. Cluster management tools deploy software and node provisioning. Cluster monitoring tools, such as *ganglia* and *nagios*, may come bundled with cluster management to monitor the usage and health of the cluster. To eliminate unnecessary data transfers, NGS instruments write data directly, via a high-bandwidth network connection, to a shared data storage system. Illumina, for example, recommends a direct 1 Gb (gigabit) network connection between the PC, dedicated to the operation of the HiSeq 2000 Sequencing System (llumina 2010), and the data storage system. Thus the collected data are immediately accessible by the HPC cluster computing nodes.

A resource manager (sometimes referred to as a workload management or simply as a batch system) is a key component of running next-generation sequencing workflows. It is the middleware used to manage jobs on the cluster by trying to match job requirements to available computing resources. User-submitted jobs are scheduled based on a scheduling policy (fair share, FIFO, etc.). Resource managers group available resources into overlapping queues based on common features (such as the high-memory queue that includes nodes with large RAM, or the long queue where jobs can run for a long time). There are a variety of batch systems (several of which allow the customization of the scheduler module) ranging from enterprise-grade, to freely available, even custom-developed batch systems, supporting a variety of features that include support for parallel jobs (usually MPI-based); checkpointing (saving the current state of a job so that in the event of a system crash the only lost computation will be from the point of the last checkpoint); process migration (the ability to move one job from one node to another without the need to restart the job); heterogeneous cluster support (a cluster consisting of systems with a variety of computer architectures, resources, and operating systems); support for multiple clusters: ability to logically bind distributed clusters into a single shared resource; and cloud adaptivity (ability to dynamically provision nodes from external providers based on changing workload demands as needed to satisfy peak demand). Resource managers monitor individual jobs, can send notifications and alerts regarding a job's status, and keep extensive logs. Such logs can be processed for job accounting purposes. Some systems provide utilities that scan logs to extract accounting information of jobs per user or jobs per user group, total number of node-hours used, and the like. Logs can also be sent to the vendors or used for troubleshooting.

The usual approach is that users access the front-end node of the cluster via the secure shell (`ssh`), using a two-factor authentication based on `ssh` keys. The two-factor authentication is based on what the user knows (a pass phrase) and what the user has (a private key). To run a job on the cluster, users write a script including flags that, in addition to the application that needs to be executed, provide specific instructions to the batch system, such as sending an e-mail notification when the job starts and when it ends, requesting specific resources (such as a minimum amount of memory available on the computing node), and specifying a particular batch system queue. Usually the data are staged on file systems that are available on all nodes of the cluster via NFS. However, the user may choose to stage input or reference data on a node's local "scratch" disk before the job starts in order to achieve better IO performance and stage data out when the job completes. This conserves network bandwidth and improves IO performance by staging data on local disks on the worker nodes. In such cases, the staging of files is a task that is separate from the workflow and is incorporated in the job submission script. The user job is submitted via a simple command (such as a `qsub` command), and the scheduler will try to schedule the user job based on the scheduling policy. The user is able to check the status of the job with a command such as `qstat`.

Programmable general purpose graphics processing units (GPGPUs) provide a powerful, energy- and cost-efficient alternative to traditional HPC architectures. Graphics chips started as fixed function PC graphics cards, but because of their speed, memory bandwidth, and cost characteristics, they were soon exploited for nongraphics applications, as general purpose computing platforms for scientific and engineering computing (see, e.g., Egri et al. 2007). The introduction of CUDA, NVIDIA's software and hardware architecture that enables NVIDIA GPUs to accelerate applications written in high-level programming languages, facilitated the use of GPUs in general purpose computing, where many applications run faster than on multicore CPU systems. Although CPUs contain a few fast, high-functionality cores, GPUs contain many (several hundreds) basic cores. In addition, GPUs offer fast memory access. Thus, highly parallel applications that can use large numbers of processors can see a significant performance improvement over using standard CPUs. In the GPU programming model, GPUs work together with multicore CPUs in a hybrid architecture in which CPU cores (the "host") execute the serial part of the application, while the computationally demanding parts are offloaded and accelerated by the GPU. The CUDA architecture consists of a large number of processors grouped together into microprocessors, with several levels of memory. The global device memory can be accessed by all processors but has a relative high latency, whereas each microprocessor has small amounts of memory that are shared by all threads. Shared memory is orders of magnitude faster than device memory. In addition, each active thread is allocated several private registers that cannot be shared between threads.

Several bioinformatics tools report significant acceleration in the computational-intensive parts of pipelines when GPU-enabled. At BGI, a GPU-enabled version of SOAP, SOAP3, aligns reads against a reference DNA sequence with a 10- to 16-fold speedup when compared with the CPU-only version. GSNP, a GPU-enabled version of SOAPsnp, reports a factor of 7 improvement. GPU-BLAST (Vouzis and Sahinidis 2011) reports a speedup of threefold to fourfold. Parallel-META, a GPU and multicore CPU-based pipeline that analyzes genomic information from microbial communities, reports a 15 times acceleration when compared with traditional **metagenomic** data analysis methods (Su et al. 2011). A list of GPU-enabled bioinformatics tools is available at http://www.nvidia.com/object/bio_info_life_sciences.html.

Bio Workbench provides a list of GPU ported bioinformatics tools that can be deployed even on desktop workstations. Several cloud vendors offer GPU instances, enabling GPU computing over the Internet. Cloud vendors that provide GPU instances are listed on NVIDIA's website.

Although it is easy to find access to servers with GPU cards, GPUs are not easy to program. The required libraries to deploy such tools are relatively easily installed and configured, but a GPU porting requires serious development effort. GPU programming has many challenges. The communication between the host CPU and GPU device is slow because it uses the PCI Express bus. This makes it hard to push and get data from GPU through the PCI Express bus. Internode network bandwidth may also be an issue. Unlike CPUs, GPUs do not contain a large memory cache. Amdahl's law also applies to GPUs. Some applications that involve intense matrix calculations and highly parallel numerical tasks have reported a several-fold performance boost when compared with CPU-only implementation. Thus, porting and novel development of codes for GPUs are typically limited to large informatics centers with dedicated GPU programming experts. Fortunately, several commonly used bioinformatics packages have already been ported to GPUs and made freely available, such as BarraCUDA, SOAP3, GPU-BLAST, and others.

With the introduction of easy ways to program GPUs, advanced HPC architectures now often use heterogeneous systems in which CPUs still execute serial parts but massively parallel parts are executed on GPUs. Currently some of the most powerful clusters consist of nodes with GPU cards. Three of the top five most powerful supercomputers listed on the November 2011 edition of the top500 list (http://www.top500.org/) are equipped with GPUs: Tianhe-1A (2.57 PetaFLOPS, ranked second on the list, installed in China), Dawning (1.27, fourth, China) and TSUBAME 2.0 (1.19, fifth, Japan). NG sequencing pipelines have been ported on the above supercomputers, as, for example, described in Yutaka Akiyama (GPU Technology Conference Asia 2011), where a metagenomic analysis pipeline has been ported on TSUBAME2.0 and 60 million reads (75 bp) were analyzed per hour on 2520 GPUs.

Although many sequencing laboratories have deployed powerful computing resources, there are times when there is an urgent need to use a large amount of computing power for a short period of time ("burst" mode) that may not be available locally. For example, when a sequencer run completes or when a grant proposal submission deadline is approaching, locally available resources may be tied up, and jobs that need to execute urgently may be stuck behind long batch queues. Some laboratories, trying to account for such situations, end up overprovisioning nodes, resulting in excess cost for resources that remain idle most of the time. In such situations, the answer may be cloud computing, where resources become available and paid for as needed. Several vendors operate large data centers, where they have deployed large computing and data storage resources and are able to rent them at competitive prices using economies of scale and the flexibility of virtualization. Virtualization is what allows a physical resource, like a server, to be subdivided and host several discrete virtual servers, each using part of the available physical components on the host, in a reliable way, with negligible performance overhead, making the server appear greater than it really is (Babcock 2010). Users of cloud resources only pay for the resources they use (pay-as-you-go model). Once the resources are not needed, they are released. Vendors offer resources at prices that are competitive to what users would otherwise have to spend to buy and maintain (electricity, cooling, data center space, etc.) a similar facility locally at their institutions. The user avoids the risk of overpaying for unneeded computing resources.

Remote users, who need urgent access to resources to execute computationally intensive bioinformatics pipelines or provide a service over the Internet, can access and provision the resources they need through user-friendly web interfaces on one of the cloud vendors, such as Amazon Web Services (AWS), Microsoft Azure, or Rackspace. A typical first step is to choose one of the vendor's preexisting *machine images*, preferably one with bioinformatics tools installed. Once the virtual machine has been instantiated, it behaves and can be accessed as if it were a physical server. One of the advantages of the cloud paradigm is its elasticity; users can elastically grow or shrink the resources they use as needed. It gives a sense of infinite available resources. There are several tools that help bioinformaticians easily deploy entire clusters of virtual machines with batch systems preinstalled. StarCluster (http://web.mit.edu/stardev/cluster/), for example, is an open source cluster computing toolkit that simplifies the process of building, configuring, and running a cluster of virtual machines on the AWS Elastic Compute Cloud (EC2).

Many bioinformatics tools are available on the cloud in a variety of forms. Tools like Galaxy may come as a Software-as-a-Service (SaaS) where users can access the tool remotely, as a service available on the Internet, without necessarily having to install it and maintain it themselves locally. Some tools come in the form of a flexible, easy-to-deploy virtual machine, such as CloVR (Angiuoli et al. 2011), whereas several bioinformatics service companies, like DNAnexus (http://dnanexus.com), have built a business model around the cloud. Several tools introduce parallelization using the MapReduce programming model that was originally developed by Google for processing large amounts of web data (Dean and Ghemawat 2004). Large data sets can be distributed across thousands of processors, with each processor running the same computation on a different data set, in parallel. The MapReduce framework provides the parallelization, data distribution, and fault tolerance, hiding the details from the user, who needs only to provide a *map* task and a *reduce* task. The *map* task takes as input a function and a set of values and produces an intermediate set of key-value pairs. Following a series of intermediate steps (such as copying, grouping, and shuffling), the *reduce* task merges together values for the same key. Hadoop is an Apache Software Foundation open source implementation of the MapReduce framework than can be deployed on clusters of computing nodes. It has been used by a large number of bioinformatics projects: CloudBurst (Schatz 2009), CloudBLAST (Matsunaga et al. 2008), Crossbow (Langmead et al. 2009), and Myrna (Langmead et al. 2011).

However, there are a number of obstacles in adapting cloud computing for genomics (Armbrust et al. 2009; Schatz et al. 2010), as discussed below.

Data Transfer

Transferring very large data sets to and from cloud resources can pose a substantial barrier. To be able to use cloud resources to analyze data, the data must first be uploaded on the cloud, and the results may need to be downloaded to users' local machines. Depending on the size of data sets and the network speed, data transfers may take substantial time, maybe even longer than the time it takes to generate the data. For some (Stein 2010), this is the largest obstacle in moving genomics to the cloud. Public Internet speeds may limit data uploads to 5–10 MB/sec. Some cloud vendors may charge for uploading and downloading data. The cost may be more than $100 per terabyte transferred. To address the cost and latency of data transfers, cloud vendors also offer to physically ship hard disks via overnight delivery services instead of uploading data over the network (a service sometimes referred to as the "Netflix for Cloud Computing"). According to Armbrust et al. (2009), as the cost/performance ratio for remote computing increases much faster than Internet bandwidth, shipping hard drive disks via overnight delivery maybe become more attractive in the future. Amazon Web Services hosts local copies of large public data sets (GenBank, 1000 Genomes, Ensembl), and thus users can mount such volumes to their instances and gain instant access to large data sets with no time or charge for data uploads. In addition, Google recently announced plans to host the Short Read Archive (SRA) on their cloud and make it publicly available.

Data Storage Costs

The cost for long-term data storage is another obstacle. Large data sets that need to stay on cloud storage for a long time will cost substantially. In certain cases, it costs more to store the data than to pay for computing resources to analyze the data. The cloud is an expensive solution for archiving data. Most cloud vendors offer online calculators that give pretty accurate estimates of the costs to upload and store data on their resources.

Data Security

Data security is another potential barrier. Although well-understood technologies for data security—such as firewalls, file system encryption, filtering and monitoring network packets, and secure multifactor authentication—are available on the cloud, many institutions prefer not to store sensitive and confidential data on public clouds. Data sets that are subject to Health Insurance Portability and Accountability Act (HIPAA) or similar regulations and are auditable legally must remain on in-house data storage systems.

Service Outages and Data Safety

Cloud vendors take many measures to eliminate single points of failure and have achieved very high uptime and service availability, which is hard to match by in-house, local data centers. Nevertheless, a recent outage in AWS S3 (Simple Storage Service) and denial of service attacks on cloud vendors have made customers take additional measures in the event of a cloud service outage.

Lack of Standardized APIs

Moving from one cloud vendor to another is not always straightforward because of lack of standards. This makes moving from one vendor to another (because of poor service, cost, or a vendor going out of business), or using services from multiple cloud vendors nontrivial.

Although cloud computing is an interesting approach to scaling analysis capacity, sequencing costs are dropping at a much faster rate than our ability to store and analyze NGS data. Regardless of how much we invest in developing local resources or on the cloud, computing costs will continue to rise as a fraction of the total cost of using NGS technology. The real solution must come from software development and improved algorithms.

ACKNOWLEDGMENTS

We thank Carole Goble and the Taverna team at the University of Manchester, United Kingdom, for reading and correcting parts of the chapter that refer to Taverna workflows. Special thanks go to Dave Clements and the Galaxy team for correcting parts of the chapter that refer to the Galaxy project, as well as Kostas Karasavvas for the comments and communications regarding the integration of Taverna workflows with the Galaxy project. Special thanks also go to Yasmine Kieso for reading and correcting parts of the chapter and to George and Pavlos Symeonidis for providing free unlimited Internet access during the summer months.

REFERENCES

Allcock W. 2005. The Globus striped gridFTP framework and server. *Proceedings of the 2005 ACM/IEEE SC/05 Conference on Supercomputing.* November 12–18, 2005, Seattle, WA, pp. 54–64. doi: 10.1109/SC.2005.72.

Angiuoli SV, Matalka M, Gussman G, Galens K, Vangala M, Riley DR, Arze C, White JR, White O, Fricke WF. 2011. CloVR: A virtual machine for automated and portable sequence analysis from the desktop using cloud computing. *BMC Bioinformatics* **12:** 356.

Armbrust M, Fox A, Griffith R, Joseph AD, Katz RH, Knowinski A, Lee G, Patterson DA, Rabkin A, Stoica I, Zaharia M. 2009. *Above the clouds: A Berkeley view of cloud computing.* Technical Report No. UCB/EECS-2009-28. Electrical Engineering and Computer Sciences, University of California, Berkeley.

Babcock C. 2010. *Management strategies for the cloud revolution: How cloud computing is transforming business and why you can't be left behind.* McGraw-Hill, New York.

Bhagat J, Tanoh F, Nzuobontane E, Laurent T, Orlowski J, Roos M, Wolstencroft K, Aleksejevs S, Stevens R, Pettifer S, et al. 2010. BioCatalogue: A universal catalogue of web services for the life sciences. *Nucleic Acids Res* **38:** W689–W694.

Blankenberg D, Von Kuster G, Coraor N, Ananda G, Lazarus R, Mangan M, Nekrutenko A, Taylor J. 2010. Galaxy: A Web-based genome analysis tool for experimentalists. *Curr Protoc Mol Biol* **89:** 19.10.1–19.10.21.

Cock PJ, Antao T, Chang JT, Chapman BA, Cox CJ, Dalke A, Friedberg I, Hamelryck T, Kauff F, Wilczynski B, de Hoon MJ. 2009. Biopython: Freely available Python tools for computational molecular biology and bioinformatics. *Bioinformatics* **25:** 1422–1423.

Cock PJ, Fields JC, Goto N, Heuer LM, Rice MP. 2010. The Sanger FASTQ file format for sequences with quality scores, and the Solexa/Illumina FASTQ variants. *Nucleic Acids Res* **38:** 1767–1771.

Dean J, Ghemawat S. 2004. MapReduce: Simplified data processing on large clusters. In *Proceedings of the 6th Symposium on Operating System Design and Implementation,* December 2004. Usenix Association, San Francisco.

Deelman E, Gannon D, Shields MS, Taylor I. 2009. Workflows and e-Science: An overview of workflow system features and capabilities. *Future Gener Comput Syst* **25:** 528–540.

Egri GI, Fodor Z, Hoelbling C, Katz SD, Nogradi D, Szabo KK. 2007. Lattice QCD as a video game. *Comput Phys Commun* **177:** 631–642.

Gentleman RC, Carey VJ, Bates DM, Bolstad B, Dettling M, Dudoit S, Ellis B, Gautier L, Ge Y,

Gentry J, et al. 2004. Bioconductor: Open software development for computational biology and bioinformatics. *Genome Biol* 5: R80. doi: 10.1186/gb-2004-5-10-r80.

Goble C, De Roure D. 2009. The impact of workflow tools on data-centric research. In *The fourth paradigm: Data-intensive scientific discovery* (ed. Hey T, et al.), Part 3, pp. 137–145. Microsoft Research, Redmond, WA.

Goble CA, Bhagat J, Aleksejevs S, Cruickshank D, Michaelides D, Newman D, Borkum M, Bechhofer S, Roos M, Li P, De Roure D. 2010. myExperiment: A repository and social network for the sharing of bioinformatics workflows. *Nucleic Acids Res* **38:** W677–W682.

Goecks J, Nekrutenko A, Taylor J; The Galaxy Team. 2010. Galaxy: A comprehensive approach for supporting accessible, reproducible, and transparent computational research in the life sciences. *Genome Biol* **11:** R86. doi: 10.1186/gb-2010-11-8-r86.

Holland RCG, Down T, Pocock M, Prlić A, Huen D, James K, Foisy S, Dräger A, Yates A, Heuer M, Schreiber MJ. 2008. BioJava: An open-source framework for bioinformatics. *Bioinformatics* **24:** 2096–2097.

Hull D, Wolstencroft K, Stevens R, Goble C, Pocock M, Li P, Oinn T. 2006. Taverna: A tool for building and running workflows of services. *Nucleic Acids Res* **34:** 729–732.

Illumina. 2010. HiSeq sequencing system: Site preparation guide. Illumina proprietary catalog # SY0940-1003, Part # 15006407. Rev D June 2010. Illumina, Inc., San Diego.

Illumina. 2011a. CASAVA v1.8 user guide, Illumina proprietary, Part # 15011196. Rev B, May 2011. Illumina, Inc., San Diego.

Illumina. 2011b. Improved accuracy for ELAND and variant calling. Technical note no. 770-2011-005. Illumina, Inc., San Diego.

Kahn SD. 2011. On the future of genomic data. *Science* **331:** 728–729.

Karasavvas K, Chichester K, Cruickshank D, Haines R, Fellows R, Roos M. 2011. Enacting Taverna workflows through Galaxy. In *12th Annual Bioinformatics Open Source Conference, BOSC 2011*, July 15–16, Vienna, Austria.

Klus P, Lam S, Lyberg D, Cheung MS, Pullan G, McFarlane I, Yeo GSH, Lam BYH. 2012. BarraCUDA—A fast short read sequence aligner using graphics processing units. *BMC Res Notes* **5:** 27.

Langmead B, Schatz MC, Lin J, Pop M, Salzberg SL. 2009. Searching for SNPs with cloud computing. *Genome Biol* **10:** R134. doi: 10.1186/gb-2009-10-11-r134.

Langmead B, Hansen K, Leek J. 2011. Cloud-scale RNA-sequencing differential expression analysis with Myrna. *Genome Biol* **11:** R83. doi: 10.1186/gb-2010-11-8-r83.

Li H, Durbin R. 2009. Fast and accurate short read alignment with Burrows–Wheeler Transform. *Bioinformatics* **25:** 1754–1760.

Li H, Handsaker B, Wysoker A, Fennell T, Ruan J, Homer N, Marth G, Abecasis G, Durbin R; 1000 Genome Project Data Processing Subgroup. 2009. The Sequence Alignment/Map (SAM) format and SAMtools. *Bioinformatics* **25:** 2078–2079.

Liu CM, Lam TW, Wong T, Wu E, Yiu SM, Li Z, Luo R, Wang B, Yu C, Chu X, et al. 2012. SOAP3: Ultra-fast GPU-based parallel alignment tool for short reads. *Bioinformatics* **28:** 878–879.

Matsunaga A, Tsugawa M, Fortes J. 2008. CloudBLAST: Combining MapReduce and virtualization on distributed resources for bioinformatics applications. *IEEE 4th International Conference on eScience*, December 1–12, pp. 222–229, Indianapolis, IN.

Morgan M, Andres S, Lawrence M, Aboyoun P, Pages H, Gentleman R. 2009. ShortRead: A Bioconductor package for input, quality assessment and exploration of high-throughput sequence data. *Bioinformatics* **25:** 2607–2608.

Romano P. 2008. Automation of in-silico data analysis processes through workflow management systems. *Brief Bioinform* **9:** 55–68.

Savage CJ, Vickers AJ. 2009. Empirical study of data sharing by authors publishing in PLoS Journals. *PLoS ONE* **4:** e7078. doi: 10.1371/journal.pone.0007078.

Schatz MC. 2009. CloudBurst: Highly sensitive read mapping with MapReduce. *Bioinformatics* **25:** 1363–1369.

Schatz MC, Langmead B, Salzberg SL. 2010. Cloud computing and the DNA data race. *Nat Biotechnol* **20:** 691–693.

Stajich JE, Block D, Boulez K, Brenner SE, Chervitz SA, Dagdigian C, Fuellen G, Gilbert JGR, Korf I, Lapp H, et al. 2002. The BioPerl toolkit: Perl modules for the life sciences. *Genome Res* **12:** 1611–1618.

Stein LD. 2010. The case for cloud computing in genome informatics. *Genome Biol* **11:** 207. doi: 10.1186/gb-2010-11-5-207.

Su X, Xu J, Ning K. 2011. Parallel-META: A high-performance computational pipeline for metagenomic data analysis. In *Proceedings of the IEEE International Conference on Systems Biology*, September 2–4, 2011, pp. 173–178.

Vouzis PD, Sahinidis NV. 2011. GPU-BLAST: Using graphics processors to accelerate protein sequence alignment. *Bioinformatics* **27:** 182–188.

WWW RESOURCES

http://biosharing.org/ BioSharing homepage.

http://cran.at.r-project.org The Comprehensive R Archive Network (CRAN) provides a collection of code and documentation mirrored on a network of web and FTP servers.

http://dnanexus.com DNAnexus, a bioinformatics service company, homepage.

http://drmaa.org/ Distributed Resource Management Application API (DRMAA) homepage.

http://fasterdata.es.net/fasterdata/say-no-to-scp/ Energy Sciences Network homepage.

http://galaxyproject.org Galaxy is an open source, web-based genomic workbench that enables users to perform integrative computational analysis of genomic data.

http://git-scm.com/ git is a fast, robust, open source version control system.

http://web.mit.edu/stardev/cluster/ StarCluster is an open source cluster computing toolkit.

http://www.biocatalogue.org/ *Bio*Catalogue provides a curated catalog of Life Science web services. The University of Manchester and the European Bioinformatics Institute (EMBL-EBI).

http://www.bioconductor.org/ Bioconductor is an open source software that provides tools for the analysis and comprehension of high-throughput genomic data.

http://www.bioconductor.org/help/bioconductor-cloud-ami/ Bioconductor in the cloud homepage.

http://www.bioconductor.org/help/workflows/high-throughput-sequencing/ Bioconductor for sequence data.

http://www.bioperl.org BioPerl is a collection of Perl modules that facilitate the development of Perl scripts for bioinformatics applications.

http://www.cpan.org Comprehensive Perl Archive Network (CPAN) homepage.

http://www.globus.org/ Globus is an open source Grid software.

http://www.myexperiment.org/ myExperiment makes it easy to find, use, and share scientific workflows and other research objects. The University of Manchester and University of Southampton.

http://www.nvidia.com/object/bio_info_life_sciences.html NVIDIA list of GPU-enabled bioinformatics tools.

http://www.r-project.org/ R is a free software environment for statistical computing and graphics hosted by WU Wein.

http://www.taverna.org.uk/ Taverna homepage. A suite of tools used to design and execute scientific workflows.

https://www.globusonline.org/ An easy to use, grid-based tool, hosted on the Amazon Web Services cloud, that can improve the speed of file transfers several times faster than *scp* and *sftp*.

https://www.xsede.org/web/guest/data-transfers Extreme Science and Engineering Discovery Environment data transfers and management page.

Glossary

16S ribosomal DNA (rDNA) (Chapter 11): The 16S rRNA is a structural component of the bacterial ribosome (part of the 30S small subunit). The 16S rDNA is the gene that encodes this RNA molecule. Owing to its essential role in protein synthesis, this gene is highly conserved across all prokaryotes. There are portions of the 16S gene that are extremely highly conserved, so that a single set of "universal" PCR primers can be used to amplify a portion of this gene from nearly all prokaryotes. The gene also contains variable regions that can be used for taxonomic identification of bacteria. Amplification and taxonomic assignment of 16S rDNA sequences is a widely used method for metagenomic analysis.

Algorithm (Chapter 2): A step-by-step method for solving a problem (a recipe). In bioinformatics, it is a set of well-defined instructions for making calculations. The algorithm can then be expressed as a set of computer instructions in any software language and implemented as a program on any computer platform.

Alignment (Chapter 3): See Sequence alignment.

Alignment algorithm (Chapter 3): See Sequence alignment.

Allele (Chapter 8): In genetics, an allele is an alternative form of a gene, such as blue versus brown eye color. However, in genome sequencing, an allele is one form of a sequence variant that occurs in any position on any chromosome, or a sequence variant on any sequence read aligned to the genome—regardless of its effect on phenotype, or even if it is in a gene. In some cases, "allele" is used interchangeably with the term *genotype*.

Amplicon (Chapter 11): An amplicon is a specific fragment or locus of DNA from a target organism (or organisms), generally 200–1000 bp in length, copied millions of times by the polymerase chain reaction (PCR). Amplicons for a single target (i.e., a

Most of these terms appear in more than one chapter. The chapter listed in parentheses after a term is the chapter in which the main discussion of the term appears.

reaction with a single pair of PCR primers) can be prepared from a mixed population of DNA templates such as HIV particles extracted from a patient's blood or total bacterial DNA isolated from a medical or an environmental sample. The resulting deep sequencing provides detailed information about the variants at the target locus across the population of different DNA templates. Amplicons produced from many different PCR primers on many different DNA samples can be combined (with the aid of multiplex barcodes) into a single DNA sequencing reaction on an NGS machine.

Assemble (Chapter 2): See sequence assembly.

Assembly (Chapter 7): See sequence assembly.

BAM file (Chapter 3): BAM is a binary sequence file format that uses BZGF compression and indexing. BAM is the binary compressed version of the SAM (Sequence Alignment/Map) format, which contains information about each sequence read in an NGS data set with respect to its alignment position on a reference genome, variants in the read versus the reference genome, mapping quality, and the sequence quality string in an ASCII string that represents PHRED quality scores.

BED file (Chapter 3): BED is an extremely simple text file format that lists positions on a reference genome with respect to chromosome ID and start and stop positions. NGS reads can be represented in BED format, but only with respect to their position on the reference genome; no information about sequence variants or base quality is stored in the BED file.

BLAST (Chapter 2): The Basic Local Alignment Search Tool was developed by Altschul and other bioinformaticans at the NCBI to provide an efficient method for scientists to use similarity-based searching to locate sequences in the GenBank database. BLAST uses a heuristic algorithm based on a hash table of the database to accelerate similarity searches, but it is not guaranteed to find the optimal alignment between any two sequences. BLAST is generally considered to be the most widely used bioinformatics software.

Burrows–Wheeler transformation (BWT) (Chapter 4): BWT is a method of indexing (and compressing) a reference genome into a graph data structure of overlapping substrings, known as a suffix tree. It requires a single computational effort to build this graph for a particular reference genome, then it can be stored and reused when mapping multiple NGS data sets to this genome. The BWT method is particularly efficient when the data contain runs of repeated sequences, as in eukaryotic genomes, because it reduces the complexity of the genome by collapsing all copies of repeated strings. BWT works well for alignment of NGS reads to a reference genome because the sequence reads generally match perfectly or with few mismatches to the reference. BWT methods work poorly when many mismatches and indels are present in the reads, because many alternate paths through the suffix tree must be mapped.

Highly cited NGS alignment software that makes use of BWT includes BWA, Bowtie, and SOAP2.

Capillary DNA sequencing (Chapter 1): This is a method used in DNA sequencing machines manufactured by Life Technologies Applied Biosystems. The technology is a modification of Sanger sequencing that contains several innovations: the use of fluorescent labeled dye terminators (or dye primers), cycle sequencing chemistry, and electrophoresis of each sample in a single capillary tube containing a polyacrylamide gel. High voltage is applied to the capillaries causing the DNA fragments produced by the cycle sequencing reaction to move through the polymer and separate by size. Fragment sizes are determined by a fluorescent detector, and the bases that comprise the sequence of each sample are called automatically.

ChIP-seq (Chapter 9): Chromatin immunoprecipitation sequencing uses NGS to identify fragments of DNA bound by specific proteins such as transcription factors and modified histone subunits. Tissue samples or cultured cells are treated with formaldehyde, which creates covalent cross-links between DNA and associated proteins. The DNA is purified and fragmented into short segments of 200–300 bp, then immunoprecipitated with a specific antibody. The cross-links are removed, and the DNA segments are sequenced on an NGS machine (usually Illumina). The sequence reads are aligned to a reference genome, and protein-binding sites are identified as sites on the genome with clusters of aligned reads.

Cloning (Chapter 1): In the context of DNA sequencing, DNA cloning refers to the isolation of a single purified fragment of DNA from the genome of a target organism and the production of millions of copies of this DNA fragment. The fragment is usually inserted into a cloning vector, such as a plasmid, to form a recombinant DNA molecule, which can then be amplified in bacterial cells. Cloning requires significant time and hands-on laboratory work and creates a bottleneck for traditional Sanger sequencing projects.

Consensus sequence (Chapter 2): When two or more DNA sequences are aligned, the overlapping portions can be combined to create a single consensus sequence. In positions where all overlapping sequences have the same base (a single column of the multiple alignment), that base becomes the consensus. Various rules may be used to generate the consensus for positions where there are disagreements among overlapping sequences. A simple majority rule uses the most common letter in the column as the consensus. Any position where there is disagreement among aligned bases can be written as the letter *N* to designate "unknown." There is also a set of IUPAC ambiguity codes (YRWSKMDVHB) that can be used to specify specific sets of different DNA bases that may occupy a single position in the consensus.

Contig (Chapter 1): A contiguous stretch of DNA sequence that is the result of assembly of multiple overlapping sequence reads into a single consensus sequence. A contig requires a complete tiling set of overlapping sequence reads spanning a genomic region without gaps.

Coverage (Chapter 1): The number of sequence reads in a sequencing project that align to positions that overlap a specific base on a target genome, or the average number of aligned reads that overlap all positions on the target genome.

de Bruijn graph (Chapter 5): This is a graph theory method for assembling a long sequence (like a genome) from overlapping fragments (like sequence reads). The de Bruijn graph is a set of unique substrings (words) of a fixed length (a *k*-mer) that contain all possible words in the data set exactly once. For genome assembly, the sequence reads are split into all possible *k*-mers, and overlapping *k*-mers are linked by edges in the graph. Reads are then mapped onto the graph of overlapping *k*-mers in a single pass, greatly reducing the computational complexity of genome assembly.

De novo assembly (Chapter 5): See De novo sequencing.

De novo sequencing (Chapter 5): The sequencing of the genome of a new, previously unsequenced organism or DNA segment. This term is also used whenever a genome (or sequence data set) is assembled by methods of sequence overlap without the use of a known reference sequence. De novo sequencing might be used for a region of a known genome that has significant mutations and/or structural variation from the reference.

Diploid (Chapter 8): A cell or organism that contains two copies of every chromosome, one inherited from each parent.

DNA fragment (Chapter 1): A small piece of DNA, often produced by a physical or chemical shearing of larger DNA molecules. NGS machines determine the sequence of many DNA fragments simultaneously.

Exon (Chapter 10): A portion of a gene that is transcribed and spliced to form the final messenger RNA (mRNA). Exons contain protein-coding sequence and untranslated upstream and downstream regions (3′ UTR and 5′ UTR). Exons are separated by introns, which are sequences that are transcribed by RNA polymerase, but spliced out after transcription and not included in the mature mRNA.

FASTA format (Chapter 3): This is a simple text format for DNA and protein sequence files developed by William Pearson in conjunction with his FASTA alignment software. The file has a single header line that begins with a ">" symbol followed by a sequence identifier. Any other text on the first line is also considered the header, and any text following the first carriage return/line feed is considered part of the sequence. Multiple sequences can be stored in the same text file by adding additional header lines and sequences after the end of the first sequence.

FASTQ file (Chapter 3): A text file format for NGS reads that contains both the DNA sequence and quality information about each base. Each sequence read is represented as a header line with a unique identifier for each sequence read and a line of DNA bases represented as text (GATC), which is very similar to the FASTA format. A second pair of lines is also present for each read, another header line and then a line with a string of ASCII symbols, equal in length to the number of bases in the read, which encode the PHRED quality score for each base.

Fragment assembly (Chapter 4): To determine the complete sequence of a genome or large DNA fragment, short sequence reads must be merged. In Sanger sequencing projects, overlaps between sequence reads are found and aligned by similarity methods, then consensus sequences are generated and used to create contigs. Eventually a complete tiling of contigs is assembled across the target DNA. In NGS, there are too many sequence reads to search for overlaps among them all (a problem with exponential complexity). Alternate algorithms have been developed for de novo assembly of NGS reads, such as de Bruijn digraphs, which map all reads to a common matrix of short *k*-mer sequences (a problem with linear complexity).

GenBank (Chapter 2): The international archive of DNA and protein sequence data maintained by the National Center for Biotechnology Information (NCBI), a division of the U.S. National Library of Medicine. GenBank is part of a larger set of online scientific databases maintained by the NCBI, which includes the PubMed online database of published scientific literature, gene expression, sequence variants, taxonomy, chemicals, human genetics, and many software tools to work with these data.

Heterozygote (Chapter 8): Humans and most other eukaryotes are diploid, meaning that they carry two copies of each chromosome in every somatic cell. Therefore, each individual carries two copies of each gene, one inherited from each parent. If the two copies of the gene are different (i.e., different alleles of that gene), then the person is said to be a heterozygote for that gene. A homozygote has two identical copies of that gene. In genome sequencing, every base of every chromosome can be considered as a separate data point; thus any single base can be genotyped as heterozygous or homozygous in that individual.

High-performance computing (HPC) (Chapter 12): High-performance computing (HPC) provides computational resources to enable work on challenging problems that are beyond the capacity and capability of desktop computing resources. Such large resources include powerful supercomputers with massive numbers of processing cores that can be used to run high-end parallel applications. HPC designs are heterogeneous, but generally include multicore processors, multiple CPUs within a single computing device or node, graphics processing units (GPUs), and multiple nodes grouped in a cluster interconnected by high-speed networking systems. The most powerful current supercomputers can perform several quadrillion (10^{15}) operations per second (petaflops). Trends for supercomputing architecture are for greater miniaturization of parallel processing units, which saves energy (and reduces heat), speeds message passing, and allows for access to data in shared memory caches.

Histone (Chapter 9): In eukaryotic cells, the DNA in chromosomes is organized and protected by wrapping around a set of scaffold proteins called **histones**. Histones are composed of six different proteins (H1, H2A, H2B, H3, H4, H5). Two copies of each histone bind together to form a spool structure. DNA winds around the histone core about 1.65 times, using a length of 147 bp to form a unit known as the **nucleosome**. Methylation and other modifications of the histone proteins affect the structure and function of DNA (epigenetics).

Human Genome Project (HGP) (Chapter 1): An international effort including 20 sequencing centers in China, France, Germany, Great Britain, Japan, and the United States, coordinated by the U.S. Department of Energy and the National Institutes of Health, to sequence the entire human genome. The effort formally began in 1990 with the allocation of funds by Congress and the development of high-resolution genetic maps of all human chromosomes. The project was formally completed in two stages, the "working draft" genome in 2000 and the "finished" genome in 2003. The 2003 version of the genome was declared to have fewer than one error per 10,000 bases (99.99% accuracy), an average contig size of >27 million bases, and to cover 99% of the gene-containing regions of all chromosomes. In addition, the HGP was responsible for large improvements in DNA sequencing technology, mapping more than 3 million human SNPs, and genome sequences for *Escherichia coli*, fruit fly, and other model organisms.

Human Microbiome Project (Chapter 11): An effort coordinated by the U.S. National Institutes of Health to profile microbes (bacteria and viruses) associated with the human body—first to inventory the microbes present at various locations inside and outside the body and the normal range of variation in healthy people, then to investigate changes in these microbial populations associated with disease.

Illumina sequencing (Chapter 1): The NGS sequencing method developed by the Solexa company, then acquired by Illumina Inc. This method uses "sequencing by synthesis" chemistry to simultaneously sequence millions of ~300-bp-long DNA template molecules. Many sample preparation protocols are supported by Illumina including whole-genome sequencing (by random shearing of genomic DNA), RNA sequencing, and sequencing of fragments captured by hybridization to specific oligonucleotide baits. Illumina has aggressively improved its system through many updates, at each stage generally providing the highest total yield and greatest yield of sequence per dollar of commercially available DNA sequencers each year, leading to a dominant share of the NGS market. Machines sold by Illumina include the Genome Analyzer (GA, GAII, GAIIx), HiSeq, and MiSeq. At various times, with various protocols, Illumina machines have produced NGS reads of 25, 36, 50, 75, 100, and 150 bp as well as paired-end reads.

Indels (Chapter 8): Insertions or deletions in one DNA sequence with respect to another. Indels may be a product of errors in DNA sequencing, the result of alignment errors, or true mutations in one sequence with respect to another—such as mutations in the DNA of one patient with respect to the reference genome. In the context of NGS, indels are detected in sequence reads after alignment to a reference genome. Indels are called in a sample (i.e., a patient's genome) after variant detection has established a high probability that the indel is present in multiple reads with adequate coverage and quality, and not the result of errors in sequencing or alignment.

Intron (Chapter 10): A portion of a gene that is spliced out of the primary transcript of a gene and not included in the final messenger RNA (mRNA). Introns separate exons, which contain the protein-coding portions of a gene.

***k*tup, *k*-tuple, or *k*-mer (Chapter 2):** A short word composed of DNA symbols (GATC) that is used as an element of an algorithm. A sequence read can be broken down into shorter segments of text (either overlapping or non-overlapping words). The length of the word is called the *k*tup size. Very fast exact matching methods can be used to find words that are shared by multiple sequence reads or between sequence reads and a reference genome. Word matching methods can use hash tables and other data structures that can be manipulated much more efficiently by computer software than sequence reads represented by long text strings.

Mate-pair sequencing (Chapter 1): Mate-pair sequencing is similar to paired-end sequencing; however, the size of the DNA fragments used as sequencing templates are much longer (1000–10,000 bp). To accommodate these long template fragments on NGS platforms such as Illumina, additional sample preparation steps are required. Linkers are added to the ends of the long fragments, then the fragments are circularized. The circular molecules are then sheared to generate new DNA fragments at an appropriate size for construction of sequencing libraries (200–300 bp). From this set of sheared fragments, only those fragments containing the added linkers are selected. These selected fragments contain both ends of the original long fragment. New primers are added to both ends, and standard paired-end sequencing is performed. The orientation of the paired sequence reads after mapping to the genome is opposite from a standard paired-end method (outward facing rather than inward facing). Mate-pair methods are particularly valuable for joining contigs in de novo sequencing and for detecting translocations and large deletions (structural variants).

Metagenomics (Chapter 11): The study of complete microbial populations in environmental and medical samples. Often conducted as a taxonomic survey using direct PCR (with universal 16S primers) of DNA extracted from environmental samples. Shotgun metagenomics sequences all DNA in these samples, then attempts both taxonomic and functional identification of genes encoded by microbial DNA.

Microarray (Chapter 10): A collection of specific oligonucleotide probes organized in a grid pattern of microscopic spots attached to a solid surface, such as a glass slide. The probes contain sequences from known genes. Microarrays are generally used to study gene expression by hybridizing labeled RNA extracted from an experimental sample to the array, and then measuring the intensity of signal in each spot. Microarrays can also be used for genotyping by creating an array of probes that match alternate alleles of specific sequence variants.

Multiple alignment (Chapter 5): A computational method that lines up, as a set of rows of text, three or more sequences (of DNA, RNA, or proteins) to maximize the identity of overlapping positions while minimizing mismatches and gaps. The resulting set of aligned sequences is also known as a multiple alignment. Multiple alignments may be used to study evolutionary information about the conservation of bases at specific positions in the same gene across different organisms or about the conservation of regulatory motifs across a set of genes. In NGS, multiple alignment methods are used to reduce a set of overlapping reads that have been mapped to a region of a reference genome by pairwise alignment, to a single consensus sequence; and also to aid in the de novo assembly of novel genomes from sets of overlapping reads created by fragment assembly methods.

Next-generation (DNA) sequencing (NGS) (Chapter 1): DNA sequencing technologies that simultaneously determine the sequence of DNA bases from many thousands (or millions) of DNA templates in a single biochemical reaction volume. Each template molecule is affixed to a solid surface in a spatially separate location, and then amplified to increase signal strength. The sequences of all templates are determined in parallel by the addition of complementary nucleotide bases to a sequencing primer coupled with signal detection from this event.

Paired-end sequencing (Chapter 1): A technology that obtains sequence reads from both ends of a DNA fragment template. The use of paired-end sequencing can greatly improve de novo sequencing applications by allowing contigs to be joined when they contain read pairs from a single template fragment, even if no reads overlap. Paired-end sequencing can also improve the mapping of reads to a reference genome in regions of repetitive DNA (and detection of sequence variants in those locations). If one read contains repetitive sequence, but the other maps to a unique genome position, then both reads can be mapped.

Paired-end read (**Chapter 1**): See Paired-end sequencing.

Phred score (**Chapter 2**): The Phred software was developed by Phil Green and coworkers working on the Human Genome Project to improve the accuracy of base calling on ABI sequencers (using fluorescent Sanger chemistry). Phred assigns a quality score to each base, which is equivalent to the probability of error for that base. The Phred score is the negative log (base 10) of the error probability; thus a base with an accuracy of 99% receives a Phred score of 20. Phred scores have been adopted as the measure of sequence quality by all NGS manufacturers, although the estimation of error probability is done in many different ways (in some cases with questionable validity).

Poisson distribution (**Chapter 1**): A random probability distribution in which the mean is equal to the variance. This distribution describes rare events that occur with equal probability across an interval of time or space. In NGS, sequence reads obtained from sheared genomic DNA are often assumed to be Poisson-distributed across the genome.

Pyrosequencing (**Chapter 1**): A method of DNA sequencing developed in 1996 by Nyrén and colleagues that directly detects the addition of each nucleotide base as a template is copied. The method detects light emitted by a chemiluminescent reaction driven by the pyrophosphate that is released as the nucleotide triphosphate is covalently linked to the growing copy strand. Each type of base is added in a separate reaction mix, but terminators are not used; thus a series of identical bases (a homopolymer) creates multiple covalent linkages and a brighter light emission. This chemistry is used in the Roche 454 sequencing machines.

Reference genome (**Chapter 4**): A curated consensus sequence for all of the DNA in the genome (all of the chromosomes) of a species of organism. Because the reference genome is created as the synthesis of a variety of different data sources, it may occasionally be updated; thus a particular instance of that reference is referred to by a version number.

Reference sequence (**Chapter 3**): The formally recognized, official sequence of a known genome, gene, or artificial DNA construct. A reference sequence is usually stored in a public database and may be referred to by an accession number or other shortcut designation, such as human genome hg19. An experimentally determined sequence produced by a NGS machine may be aligned and compared to a reference sequence (if one exists) in order to assess accuracy and to find mutations.

Repetitive DNA (Chapter 4): DNA sequences that are found in identical duplicates many times in the genome of an organism. Some repetitive DNA elements are found in genomic features such as centromeres and telomeres with important biological properties. Other repetitive elements such as transposons are similar to viruses that copy themselves into many locations on the genome. Simple sequence repeats are another type of repetitive element comprised of linear repeats of 1-, 2-, or 3-base patterns such as CAGcagCAGcag.... A short sequence read that contains only repetitive sequence may align to many different genomic locations, which creates problems with de novo assembly, mapping of sequence fragments to a reference genome, and many related applications.

Ribosomal DNA (rDNA) (Chapter 10): Genes that code for ribosomal RNA (rRNA) are present in multiple copies in the genomes of all eukaryotes. In most eukaryotes, the rDNA genes are present in identical tandem repeats that contain the coding sequences for the 18S, 5.8S, and 28S rRNA genes. In humans, a total of 300–400 rDNA repeats are located in regions on chromosomes 13, 14, 15, 21, and 22. These regions form the nucleolus. Additional tandem repeats of the coding sequence for the 5S rRNA are located separately. rRNA is a structural component of ribosomes and is not translated into protein. The rRNA genes are highly transcribed, contributing >80% of the total RNA found in cells. RNA sequencing methods generally include purification steps to remove rRNA or to enrich mRNA from protein-coding genes.

Ribosomal RNA (rRNA) (Chapter 10): See Ribosomal DNA.

RNA-seq (Chapter 10): The sequencing of cellular RNA, usually as a method to measure gene expression, but also used to detect sequence variants in transcribed genes, alternative splicing, gene fusions, and allele-specific expression. For novel genomes, RNA-seq can be used as experimental evidence to identify expressed regions (coding sequences) and map exons onto contigs and scaffolds.

Roche 454 Genome Sequencer (Chapter 1): DNA sequencers developed in 2004 by 454 Life Sciences (subsequently purchased by Roche) were the first commercially available machines that used massively parallel sequencing of many templates at once. These "next-generation sequencing" (NGS) machines increased the output (and reduced the cost) of DNA sequencing by at least three orders of magnitude over sequencing methods that used Sanger chemistry, but produced shorter sequence reads. 454 machines use beads to isolate individual template molecules and an emulsion PCR system to amplify these templates in situ, then perform the sequencing reactions in a flow cell that contains millions of tiny wells that each fits exactly one bead. 454 uses pyrosequencing chemistry, which has very few base-substitution errors, but a tendency to produce insertion/deletion errors in stretches of homopolymer DNA.

SAM/BAM (Chapter 3): See BAM file.

Sanger sequencing method (Chapter 1): The method developed by Frederick Sanger in 1975 to determine the nucleotide sequence of cloned, purified DNA fragments. The method requires that DNA be denatured into single strands, then a short oligonucleotide sequencing primer is annealed to one strand, and DNA polymerase enzyme extends the primer, adding new complementary deoxynucleotides one at a time, creating a copy of the strand. A small amount of a dideoxynucleotide is included in the reaction, which causes the polymerase to terminate, creating truncated copies. In a reaction with a single type of dideoxynucleotide, all fragments of a specific size will end with the same base. Four separate reactions containing a single dideoxynucleotide (ddG, ddA, ddT, and ddC) must be conducted, and then all four reactions are run on four adjacent lanes of a polyacrylamide gel. The actual sequence is determined from the length of the fragments, which correspond to the position where a dideoxynucleotide was incorporated.

Sequence alignment (Chapter 4): An algorithmic approach to find the best matching of consecutive letters in one sequence (text symbols that represent the polymer subunits of DNA or protein sequences) with another. Generally sequence alignment methods balance gaps with mismatches, and the relative scoring of these two features can be adjusted by the user.

Sequence assembly (Chapter 5): A computational process of finding overlaps of identical (or nearly identical) strings of letters among a set of sequence fragments and iteratively joining them together to form longer sequences.

Sequence fragment (Chapter 1): A short string of text that represents a portion of a DNA (or RNA) sequence. NGS machines produce short reads that are sequence fragments that are read from DNA fragments.

Sequence variants (Chapter 4): Differences at specific positions between two aligned sequences. Variants include single-nucleotide polymorphisms (SNPs), insertions and deletions, copy number variants, and structural rearrangements. In NGS, variants are found after alignment of sequence reads to a reference genome. A variant may be observed as a single mismatched base in a single sequence read, or it may be confirmed by variant detection software from multiple sources of data.

Sequencing by synthesis (Chapter 1): This is the term used by Illumina to describe the chemistry used in its NGS machines (Illumina Genome Analyzer, HiSeq, MiSeq). The biochemistry involves a single-stranded template molecule, a sequencing primer, and DNA polymerase, which adds nucleotides one by one to a DNA strand complementary to the template. Nucleotides are added to the templates in separate reaction mixes for each type of base (GATC), and each synthesis reaction is accompanied by the emission of light, which is detected by a camera. Each nucleotide is modified with a reversible terminator, so that only one nucleotide can be added to each template. After a cycle of four reactions adding just one G, A, T, or C base to each template, the terminators are removed so that another base can be added to all templates. This cycle of synthesis with each of the four bases and removal of terminators is repeated to achieve the desired read length.

Sequencing primer (Chapter 1): A short single-stranded oligonucleotide that is complementary to the beginning of a fragment of DNA that will be sequenced (the template). During sequencing, the primer anneals to the template DNA, then DNA polymerase enzyme adds additional nucleotides that extend the primer, forming a new strand of DNA complementary to the template molecule. DNA polymerase cannot synthesize new DNA without a primer. In traditional Sanger sequencing, the sequencing primer is complementary to the plasmid vector used for cloning; in NGS, the primer is complementary to a linker that is ligated to the ends of template DNA fragments.

Sequence read, short read (Chapter 1): When DNA sequence is obtained by any experimental method, including both Sanger and next-generation methods, the data are obtained from individual template molecules as a string of nucleotide bases (represented by the letter symbols G, A, T, C). This string of letters is called a sequence read. The length of a sequence read is determined by the technology. Sanger reads are typically 500–800 bases long, Roche 454 reads 200–400 bases, and Illumina reads may be 25–200 bases (depending on the model of machine, reagent kit, and other variables). Sequence reads produced by NGS machines are often called short reads.

SFF file (Chapter 1): Standard Flowgram Format is a file type developed by Roche 454 for the sequencing data produced by their NGS machine. The SFF file contains both sequence and quality information about each base. The format was initially proprietary, but has been standardized and made public in collaboration with the international sequence databases. SFF is a binary format and requires custom software to read it or convert it to human-readable text formats.

Shotgun sequencing (Chapter 1): A strategy for sequencing novel or unknown DNA. Many copies of the target DNA are sheared into random fragments, then primers are added to the ends of these fragments to create a sequencing library. The library is sequenced by high-throughput methods to generate a large number of DNA sequence reads that are randomly sampled from the original target. The target

DNA is reconstructed using a sequence assembly algorithm that finds overlaps between the sequence reads. This method may be applied to small sequences such as cosmid and BAC clones, or to entire genomes.

Smith–Waterman alignment (**Chapter 2**)**:** A rigorous optimal alignment method for two sequences based on dynamic programming. This method always find the optimal alignment between two sequences, but it is slow and very computationally demanding because it computes a matrix of all possible alignments with all possible gaps and mismatches. The size of this matrix increases with the square of the lengths of the sequences to be aligned, and it requires huge amounts of memory and CPU time to work with genome-sized sequences.

SOLiD sequencing (**Chapter 1**)**:** The Applied Biosystems division of Life Technologies Inc. purchased the SOLiD (Supported Oligo Ligation Detection) technology from the biotech company Agencourt Personal Genomics and released the first commercial version of this NGS machine in 2007. The technology is fundamentally different from any other Sanger or NGS method in that it uses ligation of short fluorescently labeled oligonucleotides to a sequencing primer rather than DNA polymerase to copy a DNA template. Sequences are detected 2 bases at a time, and then base calls are made based on two overlapping oligos. Raw data files use a "color space" system that is different from the base calls produced by all other sequencing systems and requires different informatics software. This system has some interesting built-in error correction algorithms but has failed to show superior overall accuracy in the hands of customers. The yield of the system is similar to that of Illumina NGS machines.

Variant detection (**Chapter 8**)**:** NGS is frequently used to identify mutations in DNA samples from individual patients or experimental organisms. Sequencing can be done at the whole-genome scale; RNA-seq, which targets expressed genes; exome capture, which targets specific exon regions captured by hybridization to probes of known sequence; or amplicons for genes or regions of interest. In all cases, sequence variants are detected by alignment of NGS reads to a reference sequence and then identification of differences between the reads and the reference. Variant detection algorithms must distinguish between random sequencing errors, differences caused by incorrect alignment, and true variants in the genome of the target organism. Various combinations of base quality scores, alignment quality scores, depth of coverage, variant allele frequency, and the presence of nearby sequence variants and indels are all used to differentiate true variants from false positives. Recent algorithms have also made use of machine learning methods based on training sets of genotype data or large sets of samples from different patients/organisms that are sequenced in parallel with the same sample preparation methods on the same NGS machines.

Index

Page references followed by *f* denote figures. Page references followed by *t* denote tables.

D

E

T